Water-Soluble and Insoluble Polymers and Biopolymers for Biomedical, Environmental, and Biological Applications

Water-Soluble and Insoluble Polymers and Biopolymers for Biomedical, Environmental, and Biological Applications

Editors

Florian J. Stadler
Alberto García-Peñas

MDPI • Basel • Beijing • Wuhan • Barcelona • Belgrade • Manchester • Tokyo • Cluj • Tianjin

Editors

Florian J. Stadler
College of Materials Science and Engineering
Shenzhen University
Shenzhen
China

Alberto García-Peñas
Materials Science and Engineering and Chemical Engineering
Universidad Carlos III de Madrid
Leganés
Spain

Editorial Office
MDPI
St. Alban-Anlage 66
4052 Basel, Switzerland

This is a reprint of articles from the Special Issue published online in the open access journal *Polymers* (ISSN 2073-4360) (available at: www.mdpi.com/journal/polymers/special_issues/water_soluble_insoluble_polym_biopolym_biomed_environ_biol_appl).

For citation purposes, cite each article independently as indicated on the article page online and as indicated below:

LastName, A.A.; LastName, B.B.; LastName, C.C. Article Title. *Journal Name* **Year**, *Volume Number*, Page Range.

ISBN 978-3-0365-4220-1 (Hbk)
ISBN 978-3-0365-4219-5 (PDF)

Contents

About the Editors

Florian J. Stadler

Prof. Florian J. Stadler has been working for Shenzhen University, Shenzhen, China, as a distinguished professor since 2014. He obtained his Ph.D. from Friedrich-Alexander University Erlangen-Nürnberg, Germany. He previously worked as a Postdoc at Université Catholique de Louvain, Belgium, and Chonbuk National University, School of Semiconductor and Chemical Engineering, the Republic of Korea as a regular professor. Since 2001, he has been involved in research on polymer science, mainly focusing on structure-property relationships in various systems, ranging from polyolefins and ionomers to hydrogels and supramolecular assemblies, as well as special nanoparticles.

Alberto García-Peñas

Alberto García-Peñas has conducted intense research in different research centers and universities, such as the Institute of Materials Science of Madrid (Madrid, Spain), the Institute of Polymer Science and Technology (Madrid, Spain), the University of Lisbon (Lisbon, Portugal), the University of Karlsruhe (Karlsruhe, Germany), Shenzhen University (Shenzhen, China), the Fudan University (Shanghai, China), and the University Carlos III of Madrid (Leganés, Spain).

Dr. García-Peñas has received 10 awards and distinctions, including the world "Borealis Student Innovation Award", the recognition of the Spanish National Research Council for his doctoral thesis, and the awards from the specialized groups of the Real Sociedad Española de Química and the Real Sociedad Española de Física for his work in the field of correlation between microstructural features and final properties of polymeric materials.

Presently, Alberto García-Peñas is an assistant professor in the department of Materials Science and Engineering at the University Carlos III of Madrid. Furthermore, he is the Secretary for Academic Affairs of Álvaro Alonso Barba Institute of Chemistry and Materials Technology.

Preface to "Water-Soluble and Insoluble Polymers and Biopolymers for Biomedical, Environmental, and Biological Applications"

The use of polymers in biological applications is defined by the interactions promoted between living organisms and polymeric chains, which are generally associated with the polymers'hydrophilic and hydrophobic behaviors. However, these water-friendly structures are also very useful for other applications, such as the adsorption of pollutants from sewage water. The modulation of the final properties of water-soluble and insoluble polymers tends to define the spectra of features associated with their final applications.

Florian J. Stadler and Alberto García-Peñas
Editors

polymers MDPI

Editorial

Water-Soluble and -Insoluble Polymers and Biopolymers for Biomedical, Environmental, and Biological Applications

Florian J. Stadler [1,*] and Alberto García-Peñas [1,2,*]

1 College of Materials Science and Engineering, Shenzhen Key Laboratory of Polymer Science and Technology, Guangdong Research Center for Interfacial Engineering of Functional Materials, Nanshan District Key Laboratory for Biopolymers and Safety Evaluation, Shenzhen University, Shenzhen 518055, China

2 Departamento de Ciencia e Ingeniería de Materiales e Ingeniería Química (IAAB), Universidad Carlos III de Madrid, 28911 Leganés, Spain

* Correspondence: fjstadler@szu.edu.cn (F.J.S.); alberto.garcia.penas@uc3m.es (A.G.-P.)

Citation: Stadler, F.J.; García-Peñas, A. Water-Soluble and -Insoluble Polymers and Biopolymers for Biomedical, Environmental, and Biological Applications. *Polymers* **2022**, *14*, 2386. https://doi.org/10.3390/polym14122386

Received: 6 June 2022
Accepted: 7 June 2022
Published: 13 June 2022

In this Special Issue, several papers dedicated to biomedical, environmental, and biological applications have been assembled, representing different aspects of the field. The works submitted to this Special Issue covered water-soluble biopolymers (shellac-coated and epoxy-coated microcapsules and poly-N-isopropylacrylamide-based polymers) as well as water-insoluble ones (poly(lactic acid) and chitosan). Further, synthetic but biodegradable polymers (poly(butyl cyanoacrylate) and poly(vinyl alcohol)) have been also covered.

The work of Yan et al. [1] focuses on the synthesis of microcapsules derived from melamine with epoxy or shellac coatings to cover lime (also called linden) tree wood. These microcapsules protect the wood mechanically by improving surface hardness. Furthermore, the coatings introduce self-healing to cracks in the wood, supposedly by bridging cracks and "gluing" them together.

Beletti et al. [2] produced poly(lactic acid) dispersions via an emulsion method, which is stable under refrigerator conditions for a long time. The size of the nanoparticles can be precisely tuned with a very homogeneous size distribution by choice of the synthesis composition—variation in solvent and surfactant concentrations—addition of starch, and mixing protocol. The hydrophilic–hydrophobic balance can be tuned by the amount of surfactant in the emulsion. The advantage of this approach is that castable films can be produced that are biodegradable.

Moral-Zamorano et al. [3] synthesized copolymers based on N-isopropylacrylamide (NIPAM) and dopamine methacrylamide (DMA)—called NIDO—and further functionalized them with the organometallic complex bis(cyclopentadienyl)titanium (IV) dichloride. This work is in a line of research investigations on NIDO and related polymers with respect to its physicochemical behavior [4–9] and for various applications, especially in the biomedical field [10]. The paper of Moral-Zamorano et al. [3] determined the effect of the loading of the organometallic complex on thermoresponsivity. While increasing the loading of the organometallic complex decreases the lower critical solution temperature somewhat, it leads to a very sharp change in the behavior, which only spans a transition temperature range of ca. 8 K, while for a normal unmodified NIDO, the transition spans ca. 25 K. Furthermore, this special polymer could also be used for its interactions with DNA.

Chitosan is one of the intensively researched biopolymers derived from chitin, the main structural polyamonisaccharide of insects, arachnids, and fungi. In recent years, chitin has attracted a tremendous amount of attention for biomedical [11,12], sustainable [13], and environmental applications [14]. Maliki et al. [15] summarized this development in their mini-review with respect to giving a short overview of the different sustainable development paths of this highly important biopolymer.

Keller et al. [16] prepared tailored molecular weights for poly(butyl cyanoacrylate) (PBCA) through an anionic polymerization process. For that purpose, the authors used the postulated depolymerization–repolymerization process (DPRP) in the literature, confirming its use for obtaining custom PBCA. Furthermore, it was observed that the end-capping of the PBCA chain banned the monomer release.

Alonso-López et al. [17] studied the biodegradation of poly(vinyl alcohol)-based materials in the marine environment. This is an essential task as maritime pollution by fishing nets, microplastic and other trash has affected every single coast on planet earth, including Antarctica, and has been found in almost every oceanic sediment, including those in the Mariana trench. For this reason, it would be highly desirable to replace some of the materials used as packing with new materials that degrade rapidly under marine conditions instead of creating continued pollution of the oceans for at least the next few centuries (if humanity would stop polluting them today). Alonso-López et al.'s work improved the biodegradability of poly(vinyl alcohol) by blending it with glycerol, which increased the biodegradability, albeit not enough to be a genuinely biodegradable material [17].

These six papers in this Special Issue, although not covering the full range of the vast biopolymer field, give a good overview of some important topics that are required for a sustainable future. Clearly, continuing to use enormous amounts of non-degradable synthetic polymers will spell ecological disaster, for which using biopolymers is part of the solution for preventing it. Due to their biocompatible nature, biopolymers will be suitable for in vivo applications, which have been partially investigated in this Special Issue. However, their properties are significantly more complex than traditional synthetic polymers, making an in-depth understanding of biopolymer physics necessary. In the future, we wish that the special properties of biopolymers can be used to reduce humanity's inflicted ecological damage and improve society, especially with respect to supporting the healthcare of an aging population.

Funding: This research received no external funding.

Institutional Review Board Statement: Not applicable.

Informed Consent Statement: Not applicable.

Conflicts of Interest: The authors declare no conflict of interest.

References

1. Yan, X.; Tao, Y.; Qian, X. Effect of Microcapsules with Waterborne Coating as Core Material on Properties of Coating for Tilia Europaea and Comparison with Other Microcapsules. *Polymers* **2021**, *13*, 3167. [CrossRef] [PubMed]
2. Belletti, G.; Buoso, S.; Ricci, L.; Guillem-Ortiz, A.; Aragon-Gutierrez, A.; Bortolini, O.; Bertoldo, M. Preparations of Poly(lactic acid) Dispersions in Water for Coating Applications. *Polymers* **2021**, *13*, 2767. [CrossRef] [PubMed]
3. Moral-Zamorano, M.; Quijada-Garrido, I.; San-Miguel, V.; Serrano, B.; Baselga, J.; Hashmi, S.; Stadler, F.J.; Garcia-Penas, A. Concentration Effect over Thermoresponse Derived from Organometallic Compounds of Functionalized Poly(N-isopropylacrylamide-co-dopamine Methacrylamide). *Polymers* **2021**, *13*, 3921. [CrossRef] [PubMed]
4. Du, L.; GhavamiNejad, A.; Yan, Z.C.; Biswas, C.S.; Stadler, F.J. Effect of a functional polymer on the rheology and microstructure of sodium alginate. *Carbohydr. Polym.* **2018**, *199*, 58–67. [CrossRef] [PubMed]
5. Obiweluozor, F.O.; GhavamiNejad, A.; Maharjan, B.; Kim, J.; Park, C.H.; Kim, C.S. A mussel inspired self-expandable tubular hydrogel with shape memory under NIR for potential biomedical applications. *J. Mater. Chem. B* **2017**, *5*, 5373–5379. [CrossRef] [PubMed]
6. Vatankhah-Varnoosfaderani, M.; Hashmi, S.; GhavamiNejad, A.; Stadler, F.J. Rapid self-healing and triple stimuli responsiveness of a supramolecular polymer gel based on boron–catechol interactions in a novel water-soluble mussel-inspired copolymer. *Polym. Chem.* **2014**, *5*, 512–523. [CrossRef]
7. Vatankhah-Varnoosfaderani, M.; Ghavaminejad, A.; Hashmi, S.; Stadler, F.J. Mussel-inspired pH-triggered reversible foamed multi-responsive gel–the surprising effect of water. *Chem. Commun.* **2013**, *49*, 4685–4687. [CrossRef] [PubMed]
8. Wang, Y.; García-Peñas, A.; Gómez-Ruiz, S.; Stadler, F.J. Surrounding Interactions on Phase Transition Temperature Promoted by Organometallic Complexes in Functionalized Poly(N-isopropylacrylamide-co-dopamine methacrylamide) Copolymers. *Macromol. Chem. Phys.* **2020**, *221*, 2000035. [CrossRef]
9. Garcia-Penas, A.; Biswas, C.S.; Liang, W.; Wang, Y.; Yang, P.; Stadler, F.J. Effect of Hydrophobic Interactions on Lower Critical Solution Temperature for Poly(N-isopropylacrylamide-co-dopamine Methacrylamide) Copolymers. *Polymers* **2019**, *11*, 991. [CrossRef] [PubMed]

10. Vatankhah-Varnoosfaderani, M.; GhavamiNejad, A.; Hashmi, S.; Stadler, F.J. Hydrogen bonding in aprotic solvents, a new strategy for gelation of bioinspired catecholic copolymers with N-isopropylamide. *Macromol. Rapid. Commun.* **2015**, *36*, 447–452. [CrossRef] [PubMed]
11. Amirian, J.; Zeng, Y.; Shekh, M.I.; Sharma, G.; Stadler, F.J.; Song, J.; Du, B.; Zhu, Y. In-situ crosslinked hydrogel based on amidated pectin/oxidized chitosan as potential wound dressing for skin repairing. *Carbohydr. Polym.* **2021**, *251*, 117005. [CrossRef] [PubMed]
12. Shekh, M.I.; Amirian, J.; Stadler, F.J.; Du, B.; Zhu, Y. Oxidized chitosan modified electrospun scaffolds for controllable release of acyclovir. *Int. J. Biol. Macromol.* **2020**, *151*, 787–796. [CrossRef] [PubMed]
13. Taghizadeh, M.; Taghizadeh, A.; Yazdi, M.K.; Zarrintaj, P.; Stadler, F.J.; Ramsey, J.D.; Habibzadeh, S.; Hosseini Rad, S.; Naderi, G.; Saeb, M.R.; et al. Chitosan-based inks for 3D printing and bioprinting. *Green Chem.* **2022**, *24*, 62–101. [CrossRef]
14. Sharma, G.; Naushad, M.; Kumar, A.; Kumar, A.; Ahamad, T.; Stadler, F.J. Facile fabrication of chitosan-cl-poly(AA)/$ZrPO_4$ nanocomposite for remediation of rhodamine B and antimicrobial activity. *J. King Saud Univ.-Sci.* **2020**, *32*, 1359–1365. [CrossRef]
15. Maliki, S.; Sharma, G.; Kumar, A.; Moral-Zamorano, M.; Moradi, O.; Baselga, J.; Stadler, F.J.; Garcia-Penas, A. Chitosan as a Tool for Sustainable Development: A Mini Review. *Polymers* **2022**, *14*, 1475. [CrossRef] [PubMed]
16. Keller, B.L.; Lohmann, C.A.; Kyeremateng, S.O.; Fricker, G. Synthesis and Characterization of Biodegradable Poly(butyl cyanoacrylate) for Drug Delivery Applications. *Polymers* **2022**, *14*, 998. [CrossRef] [PubMed]
17. Alonso-Lopez, O.; Lopez-Ibanez, S.; Beiras, R. Assessment of Toxicity and Biodegradability of Poly(vinyl alcohol)-Based Materials in Marine Water. *Polymers* **2021**, *13*, 3742. [CrossRef]

polymers

Review

Chitosan as a Tool for Sustainable Development: A Mini Review

Soundouss Maliki [1], Gaurav Sharma [2,3,4,*], Amit Kumar [2,3], María Moral-Zamorano [1], Omid Moradi [5], Juan Baselga [1], Florian J. Stadler [3] and Alberto García-Peñas [1,*]

1 Departamento de Ciencia e Ingeniería de Materiales e Ingeniería Química (IAAB), Universidad Carlos III de Madrid, 28911 Leganés, Spain; soundoussa.maliki@gmail.com (S.M.); mamoralz@ing.uc3m.es (M.M.-Z.); jbaselga@ing.uc3m.es (J.B.)
2 International Research Centre of Nanotechnology for Himalayan Sustainability (IRCNHS), Shoolini University, Solan 173212, India; dramitchem@gmail.com
3 College of Materials Science and Engineering, Shenzhen Key Laboratory of Polymer Science and Technology, Guangdong Research Center for Interfacial Engineering of Functional Materials, Nanshan District Key Laboratory for Biopolymers and Safety Evaluation, Shenzhen University, Shenzhen 518060, China; fjstadler@szu.edu.cn
4 School of Science and Technology, Glocal University, Saharanpur 247001, India
5 Department of Chemistry, Shahr-e-Qods Branch, Islamic Azad University, Tehran 61349, Iran; moradi.omid@gmail.com
* Correspondence: gaurav8777@gmail.com (G.S.); alberto.garcia.penas@uc3m.es (A.G.-P.)

Abstract: New developments require innovative ecofriendly materials defined by their biocompatibility, biodegradability, and versatility. For that reason, the scientific society is focused on biopolymers such as chitosan, which is the second most abundant in the world after cellulose. These new materials should show good properties in terms of sustainability, circularity, and energy consumption during industrial applications. The idea is to replace traditional raw materials with new ecofriendly materials which contribute to keeping a high production rate but also reducing its environmental impact and the costs. The chitosan shows interesting and unique properties, thus it can be used for different purposes which contributes to the design and development of sustainable novel materials. This helps in promoting sustainability through the use of chitosan and diverse materials based on it. For example, it is a good sustainable alternative for food packaging or it can be used for sustainable agriculture. The chitosan can also reduce the pollution of other industrial processes such as paper production. This mini review collects some of the most important advances for the sustainable use of chitosan for promoting circular economy. Hence, the present review focuses on different aspects of chitosan from its synthesis to multiple applications.

Keywords: chitosan; sustainable development; circular economy; biopolymers

Citation: Maliki, S.; Sharma, G.; Kumar, A.; Moral-Zamorano, M.; Moradi, O.; Baselga, J.; Stadler, F.J.; García-Peñas, A. Chitosan as a Tool for Sustainable Development: A Mini Review. *Polymers* **2022**, *14*, 1475. https://doi.org/10.3390/polym14071475

Academic Editor: Luminita Marin

Received: 31 January 2022
Accepted: 2 April 2022
Published: 5 April 2022

Publisher's Note: MDPI stays neutral with regard to jurisdictional claims in published maps and institutional affiliations.

1. Introduction: Necessity of Alternative Materials for a Circular Economy

The new regulations promoted by numerous governments are trying to take care of the environment by protecting actions and behaviors to develop a new sustainable economy. Some of the most important goals of these laws are aimed at the reduction of the excessive consumption of non-renewable raw materials, especially those derived from natural sources. The extraction and cleaning of raw materials are responsible for soil degradation, biodiversity loss, water shortages, and global warming. The use of residues as raw materials is a new concept derived from the circular economy which could definitely contribute to the reduction of the huge amounts of trash accumulated in landfills. The concept of a circular material means that a new product can be obtained from the old one which is acting as a raw material. The new product will exhibit the same properties and qualities as the previous one, i.e., materials will remain in a continuous cycle of life. In general, a huge amount of this waste is composed of plastics whose versatility and wide range of properties makes it difficult to get a competitive alternative in terms of costs. Some biopolymers being investigated by scientists and industry are biodegradable, and specifically, obtained from agricultural and food processing waste. Chitosan is one of the most studied biopolymers due to its biocompatibility, biodegradability, adhesivity, and bioactivity. Chitosan is the second most abundant biopolymer in the world after cellulose; this arouses researchers' interest in fabricated novel and sustainable materials based on it. On the other hand, its low cost also makes it a good choice of material [1]. The chitosan is used in a wide range of applications and industries, related to agriculture, pharmacy, medicine, food, or textile among others [2–6]. Nonetheless, new developments involve biomedicine, biotechnology, wastewater treatment, catalysis, packaging, or bioimaging which are essential for a new sustainable era where chitosan can provide versatility, recyclability, and low cost. The nature and properties of chitosan lend themselves to sustainability criteria, due to its biodegradability, bioactivity, or the obtaining method, but there are also some specific applications related to sustainability where the chitosan can play an important role, in terms of efficiency, yield, and cost. Probably, the most important applications of chitosan in this field are associated with wastewater treatment, absorption of pollutants, or their uses as a chelation agent, an antiviral agent, or a substitute material in the paper industry [7]. Some of these recent advances involve chitosan for the preparation of composites or functionalized materials, such as aerogels based on chitosan and soot.

Chitosan biopolymer can be functionalized by several function groups. Functionalization can be grafting, addition, coupling, crosslinking, etc. [8]. These were tested for the adsorption of dyes and other pollutants, such as naphthalene, showing interesting results [9]. The combination of chitosan with other materials such as collagen can also increase the range of its features [10]; for instance, the preparation of tailored scaffolds which allows adapting their properties to clinical demand [10].

The preparation of nanoparticles or nanocomposites also contributes to the circular economy, as a lower amount of raw materials is necessary for developing a specific application-based sustainable materials. Nanocomposites with magnesium show great activity against different pathogens developed in many plants, such as *Acidovorax oryzae* and *Rhizoctonia solani* which both are rice pathogens [11]. A greater surface area can be obtained through the production of thin films reducing the amount of raw materials and consequently the volume of waste after use, but keeping the same properties of the original films. Some of these developments can be carried out using chitosan, specifically for the food packaging [12]. This mini review collects some of the most relevant points that chitosan can offer for sustainable development. The new trends in science are focused on green chemistry and the circular economy; this manuscript collects brief goals, methods, and applications which are essential for understanding the importance of chitosan for new generations.

1.1. Chitosan as a Renewable Material

1.1.1. Chitosan as a Biomaterial

Chitosan is obtained through the deacetylation of chitin, which is one of the most abundant biomaterials after cellulose. This one is a polysaccharide which can be found in crustaceans, insects, or fungi (Table 1) [13]. Chitin is considered a linear long-chain homopolymer which is composed of N-acetyl glucosamine, and can develop three polymorphic forms known as α-, β-, and γ-chitin [14].

Commercial chitosan (Figure 1) is composed of D-glucosamine and N-acetyl glucosamine and is produced by the partial deacetylation of chitin. This reaction carries out the change of acetamido groups into amino groups. There are three kinds of this biopolymer depending on its molecular weight: low molecular weight, high molecular weight, and oligochitosans [15].

Table 1. Some of the main chitin sources and percentages [13].

Source	Percentage (%)
Shrimps	30–40%
Squids	20–40%
Krill	20–30%
Crabs	15–30%
Fungi	10–25%
Insects	5–25%
Oysters	3–6%
Clams	3–6%

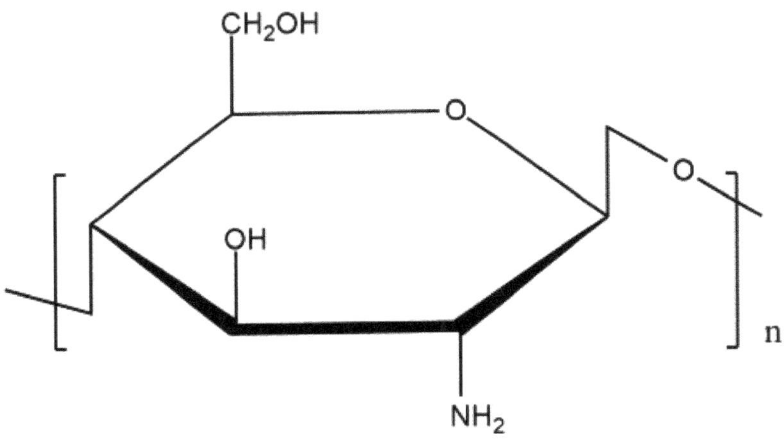

Figure 1. Chemical structure of chitosan.

1.1.2. General Features and Properties of Chitosan

The main properties which can contribute to a sustainable development that are exhibited by the chitosan are non-toxicity, biodegradability, and biocompatibility. Nevertheless, there are other interesting properties and characteristics which explain its versatility which can be deduced from Table 2.

Table 2. General properties of chitosan [16,17].

Property	Conditions	Use	References
Solubility	Dilute acids (pH < 6). Insoluble in organic solvents and water	Water treatment	[18,19]
Activity		Antibacterial, antifungal mucoadhesive analgesic, and hemostatic properties	[20–22]
Degradation	Depends on molecular weight and deacetylation degree		[18,23]
Biocompatibility	Physiological medium	Biomedical applications	[7,24]
Chelating properties	Capability to bind and adsorb diverse ions	The removal of heavy metals and dyes from wastewater	[25,26]
Biodegradability	Biodegradable to normal body constituents		[24,27,28]
Hemostatic		Stop a hemorrhage	[29,30]
Catalyst	Accelerates the formation of osteoblast		[31]
Fungicide		Stopping the development of fungi	[32,33]
Spermicidal		Reduce the mobility of spermatozoa	[34]
Anticholesteremic		Reducing agent cholesterol	[35,36]
Anticancer		Inhibiting the development of cancer cells	[37]
Conductivity	Ionic conductivity		[38,39]
Flocculating agent	Interactions with negatively charged molecules	Water treatment	[40]
Thickener		Increase the viscosity	[41]
Polyelectrolytes	Acidic medium		[42]
Adsorption		Separation and filtration	[43–45]
Clarifying agent		Immobilization of enzymes	[46]

From the presentation of Table 2, it can be deduced that chitosan is a sustainable material as it is biodegradable and non-toxicity [47]. Another important reason for using chitosan is the presence of a large number of hydroxyl and amino groups in its structure which are suitable for chemical modifications [48]. This fact and the wide versatility of chitosan makes this material especially interesting for the preparation of suspensions, composites, functionalized materials, or (nano)hybrids for diverse eco-friendly purposes and applications. The interesting polymorphic behavior exhibited by the chitosan [49], together with the molar mass and degree of deacetylation, mainly defines its mechanical properties. The molar mass will also play an important role for other properties such as degradation degree or antibacterial activity as these are strongly affected by the changes in molar mass.

On the other hand, the degree of deacetylation is associated with the content of acetamide groups of polymeric chains. These groups will strongly affect the final features and properties of the chitosan, in particular its capacity to be biodegradable and its immunological activity. The deacetylation degree is defined between 50 and 99%, its content depends on the preparation methods. The deacetylation degree must be higher than 50% for the chitosan; below that value, it is considered chitin [18]. Some of the most important uses of chitosan are associated with biomedical applications. Nevertheless, new developments related to chitosan focus on agriculture, food packaging, textiles, or environmental applications [50]. The solubility of the chitosan depends on the medium being used to dissolve it; in acid mixtures with water, it is soluble, but it is insoluble in common organic solvents [51,52]. The reason for its solubility can be explained due to the presence of amino groups that transforms chitosan into a base, whose protonation produces a polyelectrolyte [53]. The presence of different functional groups is responsible for the reactivity and the flexibility of this polycationic polymer [54]. Chitosan biofilms show a semi-crystalline behavior, together with high hydrophobicity and little flexibility [55].

1.1.3. Chitosan as an Ecofriendly Biopolymer and Its Applications

Chitosan is considered a natural biopolymer; it has received remarkable attention from the scientific community due to the fact that it can be easily biodegraded. Its residues are not toxic and can be easily eliminated and biodegraded by nature [7]. One of the most important problems associated with the raw materials is that these are limited, but chitosan is the most abundant biopolymer after cellulose. Furthermore, chitosan exhibits a great biocompatibility, limited by its low solubility which can be solved through chemical modifications and hydrolysis. Chitosan is a bioactive material which can be modulated and used in many applications [56]. Some of these applications are associated with biomedical purposes such as drug delivery systems, scaffolds, or membranes. Nevertheless, there are other important uses such as in the textile industry, wastewater treatments, agriculture, food, packaging, personal care, and biotechnology, among others. The adsorbent properties of chitosan are very useful for removing different heavy metal ions accumulated in water and derived from industrial processes such as Pb^{2+}, Hg^{2+}, and Cu^{2+}, among others [57]. These can be accumulated inside the body and produce numerous diseases [58]. Chitosan can contribute to the agriculture by improving the harvest and productivity, being an ecofriendly material. It is used as a coating for seeds, enhancing the properties of the plants and the obtained products in terms of shelf life. This use as fertilizer is especially useful for plant protection as it can stimulate the plant defense, but it can also act as an antibacterial and antimicrobial agent [59]. Thus, chitosan acts as a plant growth-promoting agent and plant protector [60]. For that reason, it is considered a pesticide by several countries. The antioxidant properties of chitosan, together with its antimicrobial features, are suitable for the production of films for food packaging. The preparation of hybrid materials with chitosan allows modifying the permeability of those films depending on the requirements [2]. The chitosan can also be used as a food additive, dietary fiber, and functional ingredient [61,62].

2. Sustainable Production

2.1. Chitin Extraction

The extraction of chitin is necessary for the production of chitosan such as it was previously explained. A huge amount of chitin is obtained from crustaceans, but there are multiple advances in its production through insects or fungi and bacteria, thus avoiding the use of animal derivatives [63]. In general, the extraction requires several steps starting with the removal of mineral salts and proteins (Figure 2). It is commonly carried out chemically, using acids and bases, which is not a sustainable process. These processes can destroy some properties of chitosan, reducing its versatility. Currently, there are multiple advances in natural deep eutectic solvents which could replace the hazardous solvents and preserve the features of chitin. There is another option based on the use of microorganisms for the extraction of chitin known as a biological method [64]. In general, these methods are especially indicated for the treatment of fungi and bacteria whilst chemical processes are related to the treatment of crustaceans. After removing the minerals and proteins, chitin requires a depigmentation process which is generally performed using oxidizing agents. The use of the enzymes could be a feasible way for removing the proteins, which can reduce the degree of depolymerization in comparison with traditional methods. That chitin also showed a better solubility in water probably due to a lower crystallinity of the product [65]. The specific use of the trypsin also induces the depigmentation, reducing the steps involved in the extraction of chitin [66]. There is a lot of ground to cover in terms of sustainability around processes for the extraction of chitin associated with environmental pollution, loss of chitin properties, and costs. One of the main consequences of this extraction is the polluted wastewater, which needs to be treated.

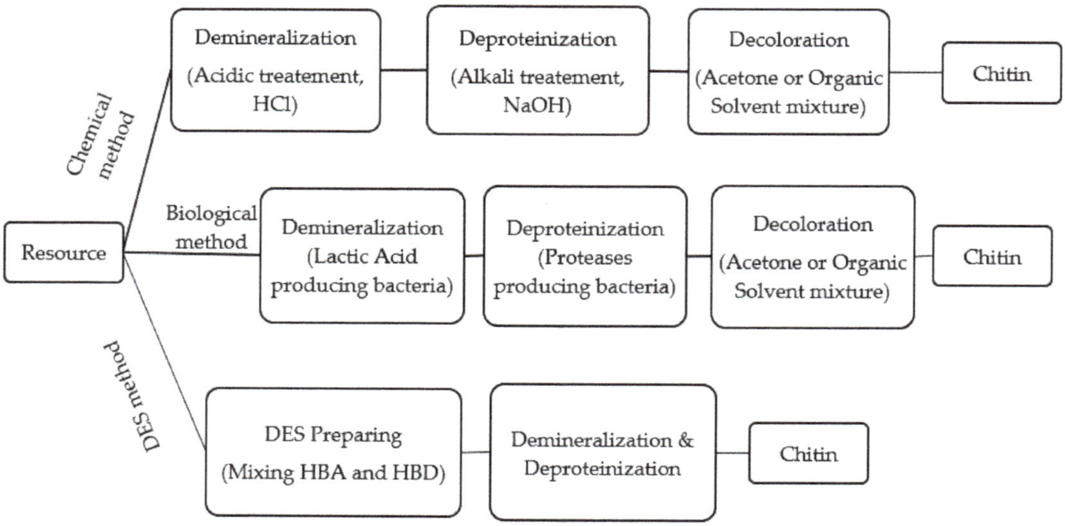

Figure 2. Extraction of chitin. DES: deep eutectic solvents; HBA: hydrogen bond acceptor; HBD: hydrogen bond donor.

2.2. Chitosan Production

The production of chitosan requires the deacetylation of chitin; this process can be modulated through concentration, temperature, and time [7]. Scheme 1 shows the changes produced in chitin after being transformed into chitosan.

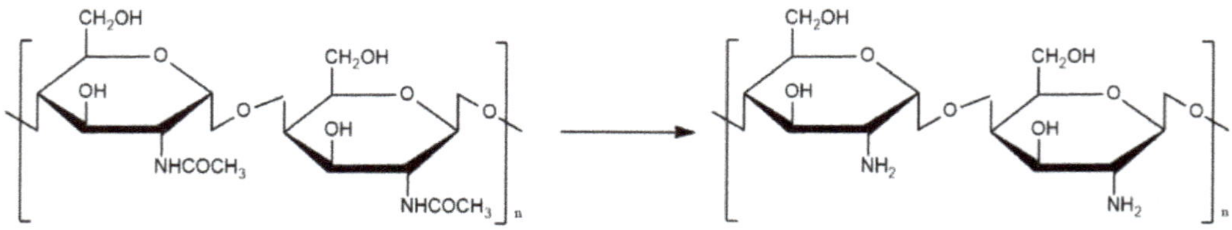

Scheme 1. Deacetylation of chitin.

The traditional method to obtain chitosan from chitin was reported in 1980, which promotes a high deacetylation due to rapid reaction rates at reduced temperatures [67]. There are different ways to carry out the deacetylation such as alkali treatment, the use of enzymes, or a steam explosion [16,68,69]. The degree of deacetylation will define the spectra of properties of the chitosan in terms of features such as solubility, viscosity, or biodegradability, etc. [70]. There are numerous alternatives where the energy consumption can be reduced, contributing to a green chemistry. Those methods explore the use of microwaves and ultrasonic waves in the deacetylation process. The use of ultrasonic waves leads to enhancing the reactivity of the deacetylation process [71]. Some of the new approaches are displayed in Table 3, showing some of the most interesting advances related to the sustainable production of chitosan.

Table 3. New methods for the production of chitosan.

Treatment	Disadvantages	Advantages	Reference
Trypsin (crustaceans)	Only for deproteination step	Depigmentation of treated material	[66]
Streptomyces griseus (crustaceans)	Only for deproteinization	Better solubility	[65]
Bacillus mojavensis A21 or Balistes capriscus (crustaceans)	Deproteinization requires NaOH	Optimized process	[72]
Rhizopus oryzae (fungi)	Fermentation	Cheap, low energy consumption, and soft conditions	[73]

2.3. *Circularity in the Chitosan Production*

The traditional methods can also be adapted, at least partially, trying to get a sustainable production of chitosan. For that purpose, it is necessary to reduce the energy consumption by reusing the hazardous reagents. The recovery of sodium hydroxide used in the extraction of chitosan was reported in studies. The sodium hydroxide is part of wastewater and could be treated using ultrafiltration and nanofiltration membranes recovering the sodium hydroxide for a new cycle of life [74,75]. The reuse of sodium hydroxide can contribute to a decrease the environmental pollution and reducing the cost of the process, i.e., a lower amount of sodium hydroxide will be required. There were also reports for the preparation of chitosan at ambient temperature, following the general procedure of demineralization, deproteinization, and decolorization [76]. This fact could also be quite interesting, due to the reduced energy consumption. Thus, involving circularity in the production of chitosan can be very beneficial and economically better.

3. Applications of Chitosan for Sustainable Development

Chitosan can contribute to sustainable development through its applications and uses. This review tries to expose some of the most important applications related to the contribution of chitosan to a circular economy and sustainability. Figure 3 depicts the diversified application of chitosan.

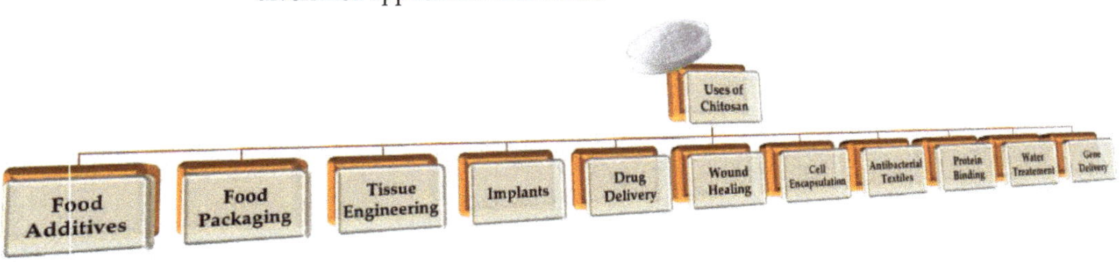

Figure 3. Different uses of chitosan.

3.1. *Sustainable Use of Chitosan for Food Packaging and in Agriculture*

Many biopolymers are being implemented in different coating materials due to their excellent properties in terms of degradability and compatibility; these biopolymers include gums, starch, proteins cellulose, lipids, and their derivatives [77–83]. In this sense, chitosan is a promising material for that purpose due to several reasons associated with its biocompatibility and abundance [84,85]. The use of the chitosan in films can also provide other superiorities because of its antibacterial and antioxidant properties [86–89]. In general, chitosan is used in combination with other polymers due to some of its drawbacks associated with its low mechanical properties. Another important problem associated with chitosan is related to its water sensitivity [90]. The preparation of blends can diminish these problems, thus obtaining films with a wide range of properties. The miscibility problems between the mixtures of polymers can reduce the spectra of possibilities, but in general, the preparation of these films is easy and cheap. The preparation of these systems could be a good alternative regarding traditional films based on oil derivatives [91]. Table 3 displays some of the most promising blends of chitosan, based on the mixtures with other biopolymers. There are other mixtures with synthetic polymer of chitosan that are not included in this review, as those do not fit the sustainability criteria of the present review. Numerous composites of chitosan have been fabricated with graphene, carbon nanotubes, activated carbon, and metal nanoparticles [92–95]. One study suggests that poly(L-lactic acid)-ZnO multilayered with cationic chitosan and anionic β-cyclodextrin can be used as a promising material in applications for the active packaging of food [96]. A novel bilayer food packing film of Ag-Metal−organic framework loaded p-coumaric acid modified chitosan (P-CS/Ag@MOF) or chitosan nanoparticles (P-CSNPs/Ag@MOF) and polyvinyl alcohol/starch (PVA/ST) was fabricated. The bilayer composite film revealed a relatively smooth surface and higher tensile strength (27.67 MPa). The P-CS/Ag@MOF bilayer films displayed better oil resistance and oxidation resistance, and the bilayer film had good UV-blocking properties and transparency [97]. The diverse blend composites of chitosan have been developed with various natural antimicrobial compounds and have been applied for antimicrobial food packaging; such antimicrobial compounds include thyme oil, spirulina, oregano essential oil, nisin, apple peel polyphenols, bamboo vinegar, cinnamon essential oil, custard apple leaves, plum peel extract, etc. [98–104]. The antibacterial nanofiber films were fabricated using gelatin, chitosan, and 3-phenyllactic acid (PLA) by electrospinning. Under acidic conditions, chitosan and PLA interacted and formed hydrogen bonds, which decreased the crystallinity of the nanofiber films. The nanofiber film had the best thermal stability, water stability, water vapor permeability, and more effective antibacterial effects against *Salmonella enterica Enteritidis* and *Staphylococcus aureus*, suggesting that the nanofiber film mat can be used as an active food packaging [105]. Similarly, Wang et al. discussed various chitosan and gelatin edible films, their synthesis strategies including casting, electrospinning, and thermoplastic method, and their properties in their review, thus highlighting importance of chitosan-based food packing films [106]. In Argentina, chitosan is produced from the waste of the shrimp industry; the synthesized chitosan has similar physicochemical properties to those of analytical grade chitosan. The chitosan coatings applied to processed lettuce at harvest increased nutritional quality and reduced microbiological contaminants in minimal processed lettuce [107]. Panda et al. fabricated ferulic acid-modified water-soluble chitosan and poly(γ-glutamic acid) polyelectrolyte multilayers films. These film surfaces possessed a reduced amount of protein adsorption; thus, these can be used as a potential good biomaterial for biomedical purposes to intensify the bio-active surface [108], thus prompting the concept of circularity and sustainability. Tables 4 and 5 show the effects of some films over the food due to the use of chitosan which could modify its properties.

Table 4. Selection of blends of chitosan with other biopolymers for food packaging.

Biopolymer	Chitosan	Characteristics	Reference
Pectin (2% *w*/*v*)	2% *w*/*v*	Good mechanical properties. Antimicrobial activity.	[109,110]
Carboxymethyl cellulose (1–2% *w*/*v*)	1% *w*/*v*	Better mechanical properties and permeability. Antioxidant and antimicrobial activity.	[111–113]
Gum arabic (1.5% *w*/*v*)	1.5% *w*/*v*	High elasticity. Antioxidant and antimicrobial activity.	[114,115]
Cassava starch (3% *w*/*v*)	0.5% *w*/*v*	Antibacterial activity.	[116]
Corn starch (5% *w*/*v*)	(1, 2, 3, and 4% *w*/*v*)	Higher tensile strength and elasticity. Lower permeability.	[117]
Rice starch (2% *w*/*v*)		Better barrier properties.	[118]

Table 5. Effects of films based on chitosan over food.

Blend	Food	Effects	References
Chitosan-glycerol film (Good mechanical and barrier properties. Stability)	Strawberry	Better preservation effect than the commercially available PE films.	[119]
Gelatin/chitosan film with nanocarriers (Fe^{III}-HMOF-5) (Good results in mechanical properties and permeability)	Apple cubes	High content of nanocarriers allows the preservation of apple cubes during 5 days.	[120]
Chitosan films (modified with mango leaf extract) (Higher hydrophobicity and tensile strength)	Cashew nuts	High oxidation resistance.	[121]
Chitosan/gelatin film with silver nanoparticles (Better hydrophobicity and antibacterial properties)	Red grapes	Antimicrobial properties and high oxidation resistance.	[122]
Polyurethane/chitosan/nano ZnO composite film (Better mechanical properties, low permeability)	Carrot	Better shelf life than polyethylene film	[19]
Pullulan/chitosan film (good barrier to O_2)	Papayas	Maintained the physiological and nutritional attributes. High shelf life.	[123]
Chitosan-TiO_2 nanocomposite film (Better tensile strength and barrier properties)	Tomatoes	Delay the ripening process and extend the storage life.	[124]
Cellulose/chitosan/polypyrrole film	Cherry tomatoes	Possess good antioxidant, antibacterial, and barrier properties	[125]
Baicalin-liposomes loaded polyvinyl alcohol-chitosan electrospinning nanofibrous films	Mushrooms	Possessed effective antibacterial properties, non-cytotoxicity, and preservation performance	[126]
Active packaging films based on chitosan and sardinella protein isolate	Shrimps	Good antioxidant and antibacterial activities	[127]

Table 5. *Cont.*

Blend	Food	Effects	References
ε-polylysine/chitosan nanofibers	Chicken	Inhibiting *Salmonella typhimurium* and *Salmonella enteritidis* on chicken	[128]
Chitosan films embedded with Apricot (*Prunus armeniaca*) oil	Bread	Better antioxidant, mechanical, and antimicrobial properties	[129]
Zein active film containing chitosan nanoparticle encapsulated with pomegranate peel extract	Pork	Addition of chitosan nanoparticle can increase the thermal stability of zein active film Film can inhibit the growth of *Listeria monocytogenes* on pork	[130]
Mahua oil-based polyurethane/chitosan/nano ZnO composite films	Carrot	Excellent anti-bacterial properties against Gram positive and Gram-negative bacteria Increase shelf life of carrot	[131]
Carboxymethyl chitosan (CMCh)-peptide conjugates	Blueberry	Extend the shelf-life of blueberry	[132]
Chitosan-based biodegradable bags	Palmer's mango	Effective in delaying ripening and preserving the quality	[133]
Composite films based on chitosan and syringic acid	Quail eggs	Films exhibited higher density, water solubility, good preservation effect	[134]
Films based on quaternary ammonium chitosan, polyvinyl alcohol, and betalains-rich cactus pears (*Opuntia ficus-indica*) extract	Shrimp	Enhanced the UV–vis light barrier, elongation-at-break, and antioxidant, antimicrobial and ammonia-sensitive properties	[135]
Chitosan coating with vacuum packaging	Beef	Extend the shelf life of beef Inhibited *S. aureus*	[136]
Chitosan coatings	Lettuce	Improve quality and extend shelf-life of minimally processed lettuce	[107]
Chitosan films incorporating litchi peel extract and titanium dioxide nanoparticles	Watercored apple	Coating treatment significantly inhibited respiration rate, weight loss, and softening	[137]
Polylactic acid/chitosan films	Indian white prawn	Antimicrobial properties	[138]
Chitosan-Gelatin (CHI-Gel) based edible coating incorporated with longkong pericarp extract (LPE)	Shrimp	Edible coating as a natural antioxidant, antimicrobial activity and inhibiting melanosis, retain the quality and extend the shelf-life	[139]
Pink pepper residue extracts incorporated in a chitosan film	Salmon fillets	Shelf-life of the skinless salmon fillet could be extended by 28 days	[140]
Chitosan film incorporated with citric acid and glycerol	Green chilies	Improved mechanical, thermal, and antioxidant properties of the film were and increased shelf life	[141]

The chitosan can act as protector, coating material, stimulator of the growth, nutrient, fertilizer, or pesticide in agriculture. It was also observed that the use of chitosan can increase productivity. Furthermore, the use of chitosan could replace some dangerous chemicals used as compounds of fertilizers in agriculture, protecting soil, aquifers, and ecosystems [142]. It was reported that excellent antimicrobial activity was observed in chitosan against many viruses, bacteria, and fungi. Nevertheless, its activity is higher against fungi than bacteria. In general, the chitosan seems to inactivate the replication of viruses [143]. Moreover, it is considered a potent elicitor which can induce plant defense against diseases [144]. Table 6 shows some of the effects observed of chitosan over some fruits and vegetables.

Table 6. Effects of chitosan and derivatives over some products.

Material/Use	Plant	Effects	Reference
Chitosan with copper	Tomato	Plant defense (Enzymatic and anatomical changes).	[145]
Seed-priming with chitosan	Cucumber	Disease protection and enhanced plant growth.	[146]
Foliar application of chitosan	Sweet pepper	Enhancement of the adverse effects of salinity and improved the growth and yield.	[147]
Chitosan solution (using a hand sprayer)	*Dracocephalum kotschyi*	Increase of antioxidant enzyme.	[148]
Chitosan (foliar spray or pre-sowing seed treatments in Cd-stressed plants)	Pea	Improvement in growth, photosynthetic pigments, and reduction in oxidative damage.	[149]
Chitosan (protective spray)	Mango (Amrapali and Dashehari)	Reduced malformation of mango.	[150]
Chitosan nanoparticles	Durum wheat	Increase the leaf antioxidant pool.	[151]
Chitosan oligosaccharide (COS)	Tea plant (*Camellia sinensis*)	Improved the antioxidant enzyme activities and the content of chlorophyll and soluble sugar.	[152]
Chitosan nanoemulsion containing allspice essential oil	Maize	Preserved maize samples from aflatoxin B1 and lipid peroxidation.	[153]
Chitosan nanoparticles loaded with garlic essential oil	Wheat, oat, and barley	As a seed dressing agent found to have antifungal activity against *Aspergillus versicolor*, *A. niger*, and *Fusarium oxysporum*.	[154]
1.5% chitosan solution treatment	Berry	Inhibit postharvest berry abscission of the 'Kyoho' table grapes.	[155]
Preharvest chitosan sprays	Muskmelons	Induced suberin polyphenolic deposition at wound sites during healing thus promoted wound healing and reduced disease development.	[156]
Chitosan film containing *Akebia trifoliata* (Thunb.) Koidz. peel extract/montmorillonite	*A. trifoliata* fruits	Significant effect on the delaying crack and mature of the fruits.	[157]
Chitosan-based nanoencapsulated *Foeniculum vulgare* Mill. essential oil	*Sorghum bicolor*	Significantly preserved the nutritional and sensory characteristics of *S. bicolor* seeds.	[158]
Encapsulated peppermint essential oil in chitosan nanoparticles	-	Biological efficacy against stored-grain pest control.	[159]

3.2. *Sustainable Applications of Chitosan in Purification of Water, Paper-Making, and Green Chemistry*

The chitosan is a good flocculant for water treatment, especially indicated for organic matter, suspended solids, and ions (metals). Furthermore, the deposition rate is stimulated when chitosan is used [160]. It is used over oil spills as it can preserve the integrity of the oil mass. Its properties are also indicated for anionic waste where the chitosan can remove the metal ions of the acid solutions. Some of the most attractive features of chitosan regarding other flocculants are associated with its biodegradability and its adsorption and flocculating ability, which show excellent results with oils [7]. However, there are many other pollutants where the chitosan shows interesting results as can be observed in Table 7. Chitosan and its composites demonstrate excellent adsorption properties for diversified environmental contaminates ranging from organic pollutants to metal ions [47,161–165]. The mechanism for the adsorption of toxic pollutants by chitosan and its composites involves various types of interactions such as electrostatic, hydrogen bonding, π-π bonding, etc. The chitosan and its composites had several hydroxyls and amino and carboxylic groups which are very helpful for such interactions, thus making it more adsorbent.

Table 7. Examples of pollutants removed by chitosan and derivatives.

Pollutant	Adsorbent	Efficiency	References
Tetracycline	Chitosan/poly (vinyl alcohol) nanofibers	102 mg/g (maximum adsorption capacity)	[166]
Ciprofloxacin	Chitosan/biochar hydrogel	36.72 mg/g (uptake capacity)	[167]
Tetracycline	Magnetic polymer nanocomposite was fabricated using chitosan, diphenyl urea, and formaldehyde	168.24 mg/g (maximum adsorption capacity)	[168]
Tetracycline	Nanocomposite of chitosan/thiobarbituric acid/malondialdehyde-Fe_3O_4	215.31 mg/g (highest adsorption capacity)	[169]
Antibiotics	Chitosan-grafted SiO_2/Fe_3O_4 nanoparticles	100.74 mg/g (theoretical adsorption capacity)	[170]
Ketoprofen	Chitosan/Zr-MOF (UiO-66) composite	Maximum adsorption capacity of 209.7 mg/g	[171]
Tetracycline	Nitrilotriacetic acid modified magnetic chitosan-based microspheres	Adsorption capacity of 373.5 mg g^{-1}	[172]
Congo red	Chitosan nanoparticles	99.96%	[173]
Methylene blue	Chitosan/κ-carrageenan/acid-activated bentonite composite membranes	Maximum adsorption capacity for methylene blue was 18.80 mg/g	[174]
Azo dyes	Glass beads coated with chitosan	Maximum adsorption capacity of the column packed with GBCC was 108.7 mg g^{-1}.	[175]
Methyl orange	Chitosan-lysozyme biocomposite	Maximum adsorption capacity for MO was 435 mg/g	[176]
Methylene blue	Bivinylbenzene cross-linked chitosan/maleic anhydride polymer	Adsorption capacity for MB 503 mg/g	[177]
Acid orange 7 (AO7, monovalent), Acid red 13 (AR13, divalent), and Acid red 27 (AR27, trivalent) dyes	Chitosan–magnetite gel microparticles	Acid Orange 7 (AO7, monovalent), Acid Red 13 (AR13, divalent), and Acid Red 27 (AR27, trivalent) dyes with maximum adsorption capacities, Q_{max}, of 1.71, 1.55, and 1.13 g-dye/g-dry adsorbent, respectively	[178]
Methyl orange dye	Fe-loaded chitosan film	Maximum adsorption capacity 205 mg g^{-1}	[179]
Methyl orange dye	Chitosan/carbon/Fe_3O_4	Maximum adsorption capacity was 425 mg g^{-1}	[180]
Disperse blue 367	Magnetic/chitosan/graphene oxide	Adsorption capacity of 298.27 mg/g	[181]
Reactive orange 16 dye	Chitosan tripolyphosphate/TiO_2 nanocomposite	Adsorption capacity was 618.7 mg/g	[182]

Table 7. *Cont.*

Pollutant	Adsorbent	Efficiency	References
Acid red 88	Phosphorylated chitosan	Adsorption capacity was 230 mg g^{-1}	[183]
Methylene blue	Poly(glycerol sebacate)/chitosan/graphene oxide nanocomposites	Adsorption capacity was 129 mg/g	[184]
Methylene blue	Magnetic sodium ferrosilicate/carboxymethyl chitosan composite	Adsorption capacity was 515.0 mg/g	[185]
Malachite green (MG), reactive red (RR), and direct yellow (DY) dyes	Chitosan	Adsorption capacities 166 mg/g for dye MG, 1250 mg/g for dye RR and 250 mg/g for dye DY	[186]
Methyl orange	Chitosan crosslinked with metal-organic framework (MOF-199)@aminated graphene oxide aerogel	Maximum adsorption capacity for methyl orange 412 mg/g	[187]
Reactive orange 16	Chitosan-polyvinyl alcohol/fly ash (m-Cs-PVA/FA)	Adsorption capacity of m-Cs-PVA/FA for RO16 dye removal was 123.8 mg/g	[188]
Methyl orange and methylene blue	Graphene oxide-chitosan composite	Maximum adsorption amounts of MO and MB were 543.4 and 110.9 mg/g	[189]
Phenol, BPA, and 2,4-DCP	Chitosan modified nitrogen-doped porous carbon composite	Maximum adsorption capacity for phenol, BPA, and 2,4-DCP was 254.45, 675.68, and 892.86 mg g^{-1}	[190]
Sunset yellow	Chitosan	Maximum adsorption capacity 1432.98 mg g^{-1}	[191]
Allura red	Luffa-chitosan crosslinked with glutaraldehyde (LCsG) and epichlorohydrin (LCsE)	LCsG and LCsE presented maximum capacities of 89.05 mg/g and 60.91 mg/g.	[192]
Brilliant blue	Chitosan	Maximum adsorption capacity 814.27 mg/g	[191]
Tartrazine	Chitosan	Maximum adsorption capacity 1065.55 mg/g	[191]
Acid blue-25	Chitosan/porous carbon composite modified in 1-allyl-3-methyl imidazolium bromide ionic liquid	Maximum adsorption capacity 3333.33 mg/g	[193]
Morphine, codeine, ephedrine, amphetamine, and benzoylecgonine	Magnetic chitosan-graphene oxide-ionic liquid ternary nanohybrid	Adsorption capacity for morphine, codeine, ephedrine, amphetamine, and benzoylecgonine (7.2, 8.4, 9.2, 5.8, and 11.2 mg g−1, respectively)	[194]
Tartrazine	Chitosan/polyaniline composite	Maximum adsorption capacity of 584.0 mg/g	[195]
Acetaminophen	Polyaniline with chitosan	Adsorption rate of 385.25 mg.g^{-1}	[196]
Anthocyanins	Chitosan beads	Adsorption capacity was 216 mg g^{-1}	[197]
Tetracycline	Zirconium-loaded chitosan modified by perlite (Zr/Cht/Pt) composites	Maximum adsorption capacity of 104.17 mg/g	[198]
Levofloxacin, tetracycline hydrochloride, and sulfamethoxazole	Chitosan	Adsorption capacity of levofloxacin, tetracycline hydrochloride, and sulfamethoxazole were 26, 22, and 67 mg/g	[199]
17α-ethinylestradiol	Graphene oxide, magnetic chitosan, and organophilic clay composite	Maximum adsorption capacity was 50.5 mg/g	[200]
Tartrazine	Surfactant-ionic liquid bi-functionalization of chitosan beads	Adsorption capacity was found to be 45.95 mg/g	[201]

The chitosan also showed good results associated with ions, as it can be observed in Table 8. These are only some examples of the good results that can be achieved.

Table 8. Examples of chitosan for removing ions.

Ion	Adsorbent	Efficiency	References
Cr (VI), Cu (II), and Co (II)	Polyethylenimine-grafted chitosan electrospun membrane	138.96, 69.27, and 68.31 mg/g for Cr(VI), Cu(II), and Co(II), respectively (maximum adsorption capacities)	[202]
Cu^{2+} and Cr^{6+}	Zeolitic imidazolate framework-67 modified bacterial cellulose/chitosan composite aerogel	200.6 mg/g and 152.1 mg/g, for Cu^{2+} and Cr^{6+}, respectively (adsorption capacities)	[203]
Cu^{2+}	Monodispersed chitosan microspheres	75.52 mg/g (adsorption capacity)	[204]
Pb^{2+}, Cu^{2+}, and Cd^{2+}	Physically crosslinked chitosan/sodium alginate/calcium ion double-network hydrogel	176.50 mg/g, 70.83 mg/g, and 81.25 mg/g for Pb^{2+}, Cu^{2+}, and Cd^{2+}, respectively (adsorption capacities)	[205]
Cu^{2+}, Pb^{2+}, and Cd^{2+}	Chitosan-coated argillaceous limestone	64.11 mg/g, 217.4 mg/g, and 52.48 mg/g for Cu^{2+}, Pb^{2+} and Cd^{2}, respectively (maximum adsorption capacities)	[206]
Cr(VI)	Terylene carbon-dots modified chitosan non-woven fabrics	Maximum adsorption capacity was 203 mg/g	[207]
Pb^{2+}	Zeolitic imidazolate framework-8 (ZIF-8) on carboxymethyl chitosan beads	Maximum adsorption capacity of 566.09 mg/g	[208]
Cd^{2+}	Cellulose/chitosan composite spheres loaded with nZVI	Maximum adsorption up to 110.3 mg/g	[209]
Cu^{2+} and Ni^{2+}	Tripolyphosphate-crosslinked-chitosan-modified montmorillonite	Adsorption capacity for Cu^{2+} and Ni^{2+} 0.56 and 0.44 mmol/g	[210]
Cr^{4+}	Chitosan-lysozyme biocomposite	Maximum adsorption 216 mg g^{-1}	[176]
Pb^{2+} and Cd^{2+}	Chitosan/Mg-Al-layered double hydroxide nanocomposite	Maximum capacities were 333.3 mg/g for Pb^{2+} and 140.8 mg/g for Cd2^{+}, respectively.	[211]
Arsenic	Silica-stabilized magnetic chitosan Beads	Maximum adsorption capacity 1.699 mg/g	[212]
Cr(III) and Cr(VI)	Iron oxide/carbon nanotubes/chitosan magnetic composite film	Maximum adsorption capacity for Cr(III) of 66.25 mg/g and for Cr(VI) of 449.30 mg/g	[213]
Cu(II)	Chitosan-coated magnetic nanoparticles	Maximum adsorption capacity was found to be 236.7 mg/g	[214]
Cr(VI)	Nano-graphene oxide-assisted hydrotalcite/chitosan biocomposite	Maximum adsorption capacity of 42.64 mg/g	[215]
Pb^{2+} and Hg^{2+}	Schiff base based on porous chitosan-glutaraldehyde/montmorrilonite nanoparticles modified with 3-aminopropyl triethoxysilane	Maximum adsorption capacity of Pb^{2+} and Hg^{2+} were 32.786 and 30.395 mg/g	[216]
Re(VII)	Chitosan-silica composite containing Mo-imprinted cavities	Adsorption capacity of 368.8 mg g^{-1}	[217]
Uranium	Chitosan-grafted adenosine 5′-monophosphate foam	Adsorption capacity of 311 mg/g	[218]

Table 8. *Cont.*

Ion	Adsorbent	Efficiency	References
Li^+	$H_4Mn_5O_{12}$/chitosan	Adsorption capacity reached 11.4 mg/g	[219]
Fluoride	Zirconium (IV)-impregnated magnetic chitosan graphene oxide	Adsorption capacity was 8.84 mg/g	[220]
U(VI)	Chitosan-based aerogel	U(VI) adsorption capacity of 160 mg/g	[221]
Au(III)	Chitosan functionalized with N,N-(2-aminoethyl)pyridinedicarboxamide	Maximum adsorption capacity of 659.02 mg/g	[222]
Cr(IV)	Chitosan composite	Adsorption capacity was 18 mg/g	[223]
Cu(II)	Benzothiazole functionalized chitosan	Maximum copper adsorption capacity of 1439.7 mg/g	[224]
Cr(IV)	Chitosan-crosslinked-poly(alginic acid)	Maximum adsorption capacity 26.49 mg/g	[225]
Pb(II)	Ninhydrin-functionalized chitosan	Maximum adsorption capacity of 196 mg/g Pb(II) ions	[226]
Co^{2+} and Sr^{2+}	Fibrous chitosan biosorbent	Adsorption capacity of fibrous chitosan for Co^{2+} and Sr^{2+} was 31.3 mg g^{-1} and 20.0 mg g^{-1}	[227]
Au(III)	Benzothiazole-modified chitosan	Maximum adsorption capacity of 1072.22 mg/g	[228]
Cu(II)	Polyacrylamide-modified kaolin enhances adsorption of sodium alginate/carboxymethyl chitosan hydrogel beads	Adsorption capacity of the adsorbent was 5.5157 mg/g	[229]
Ag(I)	Chitosan-coated magnetic silica core-shell nanoparticles	126.74 mg/g	[230]
Cu^{2+}, Fe^{3+} and Pb^{2+}	Chitosan	Maximum adsorption capacity Cu^{2+}, Fe^{3+}, and Pb^{2+} were 462 270 mg/g, 934 mg/g	[199]
Sr^{2+}	Carboxymethyl chitosan gel	Maximum adsorption capacity can reach 144.73 mg/g	[231]
As(III)	MnO_2-strengthened WTRs-chitosan beads	Adsorption capacity of 36.911 mg/g	[232]
As(III), Cd(II), Cu(II), and Pb(II)	Chitosan bead-supported $MnFe_2O_4$ nanoparticles	As(III), Cd(II), Cu(II), and Pb(II) was achieved maximum adsorption capacities of 9.90, 9.73, 43.94, and 11.98 mg/g	[233]

Chitosan can be used for paper manufacture due to its mechanical properties which can provide better resistance to recycled paper, reducing the consumption of chemical additives [234]. Table 9 displays the various roles of chitosan in paper production.

Table 9. Effects of chitosan in paper production.

Material/Use	Paper Application	Effects	Reference
Nanoparticles with chitosan and starch	Old corrugated containerboard (OCC)	Increase tensile and burst strength Decrease tear resistance	[235]
Chitosan and cellulose nanofibers	Paper recycling (decolorization)	Remove water-based inks	[236]
Microparticules with chitosan and bentonite	Paper reinforcement	Chitosan is a good dry strength additive	[237]
Chitosan as additive	Papermaking (aging stability of paper)	Increase tensile strength. Decrease the hydrophilicity of paper	[238]
Chitosan with zeolite as filler	Papermaking	Improve the mechanical properties of paper	
Chitosan as additive	Paper reinforcement (Kenaf paper (*Hibiscus cannabinus*))	Give a good mechanical and dry strength properties	[239]
Graphene ink from the exfoliation of graphite in pullulan, chitosan, and alginate	For strain-sensitive paper	Paper-based strain sensor, the chitosan-graphene has the best resistivity value and demonstrates the highest sensitivity towards strain	[240]

The chitosan can also be used as amino-functionalized structures for CO_2 capture. Many industrial processes could reduce their emissions using these systems. Furthermore, there are many other options where chitosan can be used to reduce the greenhouse gas emissions [241]. Table 10 displays the chitosan-based materials used for gas capture.

Table 10. Chitosan-based materials used for gas capture.

Adsorbate	Adsorbent	Effects	References
Carbon dioxide	Composite with chitosan and clay	Adsorption capacity of 344.98 mg/g	[242]
Carbon dioxide	Arginine-containing chitosan-graphene oxide aerogels	CO_2 gas adsorption was equal to 24.15 wt% (5.48 mmol g^{-1})	[243]
Palladium (II) and platinum (IV)	Cross-linked chitosan	340.3 mg/g and 203.9 mg/g for Pd and Pt, respectively (adsorption capacity)	[244]
Carbon dioxide (separation)	Membrane with carboxymethyl chitosan and carbon nanotubes	Good CO_2 selectivity and permeability	[245]
Carbon dioxide	Acetic acid-mediated chitosan	368 mg/g adsorption capacity Good CO_2 Selectivity	[246]
Carbon dioxide	Chitosan as a porosity agent	280.5 mg/g adsorption capacity	[247]
Formaldehyde gas	Chitosan crosslinked with metal-organic framework (MOF-199)@aminated graphene oxide aerogel	197.89 mg/g adsorption capacity	[187]
Carbon dioxide	Chitosan-grafted multi-walled carbon nanotubes	CO_2 uptake capacity was found to be significantly higher (1.92 ccg^{-1})	[248]

4. Future Perspectives

It is expected that chitosan uses will increase replacing other traditional materials due to its interesting properties and functionalities, but also due to it being abundant, it can be extracted using green chemistry and easily treated as waste. For these reasons, chitosan is considered a rich renewable resource where some of its shortcomings associated with solubility, mechanical properties, and porosity are being addressed due to the potential of this source.

This article shows some of the most prominent fields where chitosan is an interesting alternative to other conventional materials, but its properties will be reflected soon in other many fields due to its versatility and properties. Some of the most promising applications could be associated with specific areas such as medicine, food packaging, or biotechnology, among others.

There is a lot of room to grow in terms of the production of chitosan, the current goal of which is clearly focused on the removal of hazardous solvents and reducing the energy consumption. On the other hand, chitosan can contribute to sustainability in terms of recycling and waste management due to its degradability.

5. Conclusions

Chitosan shows an interesting range of properties which make it very useful for sustainable development due to it being abundant, biodegradable, biocompatible, and versatile. The production of chitosan is improving in terms of green chemistry, due to the hazardous chemicals being replaced by eutectic solvents, lower energy consumption has been achieved, and circularity can be applied to secondary processes. The use of chitosan in films for food packaging shows better properties than traditional films composed of polyethylene. The edible food packing with enhanced antimicrobial activity can be developed using chitosan. Numerous blends of chitosan have been developed with various essential oils and extracts which are excellent antibacterial and antifungal agents. On the other hand, the chitosan provides interesting and multiple features for a sustainable agriculture, such as a protection for the plant and increasing the production. Finally, the chitosan can contribute to green chemistry in multiple processes such as the paper industry or the treatment of wastewater, reducing the impact and contributing to the circularity of industrial processes. The chitosan-based composites, hydrogels, and membranes can be used for the remediation of diversified pollutants including dyes, antibiotics, phenols, metal ions, etc. Thus, being a second abundant biopolymer in nature, chitosan can be a potential sustainable future material.

Author Contributions: Conceptualization, A.G.-P. and G.S.; methodology, S.M. and M.M.-Z.; investigation, A.G.-P. and G.S.; resources, J.B.; writing—original draft preparation, A.G.-P. and G.S.; writing—review and editing, A.G.-P., G.S., A.K., O.M. and F.J.S.; supervision, A.G.-P. and G.S.; funding acquisition, J.B. All authors have read and agreed to the published version of the manuscript.

Funding: Author wants to thank the Erasmus+ KA107 scholarship.

Conflicts of Interest: The authors declare no conflict of interest.

References

1. Ahmed, M.E.; Mohamed, H.M.; Mohamed, M.I.; Kandile, N.G. Sustainable antimicrobial modified chitosan and its nanoparticles hydrogels: Synthesis and characterization. *Int. J. Biol. Macromol.* **2020**, *162*, 1388–1397. [CrossRef] [PubMed]
2. Manigandan, V.; Karthik, R.; Ramachandran, S.; Rajagopal, S. Chapter 15-Chitosan Applications in Food Industry. In *Biopolymers for Food Design*; Grumezescu, A.M., Holban, A.M., Eds.; Academic Press: Cambridge, MA, USA, 2018; pp. 469–491.
3. Ke, C.-L.; Deng, F.-S.; Chuang, C.-Y.; Lin, C.-H. Antimicrobial Actions and Applications of Chitosan. *Polymers* **2021**, *13*, 904. [CrossRef] [PubMed]
4. Pathania, D.; Gupta, D.; Kothiyal, N.C.; Sharma, G.; Eldesoky, G.E.; Naushad, M. Preparation of a novel chitosan-g-poly(acrylamide)/Zn nanocomposite hydrogel and its applications for controlled drug delivery of ofloxacin. *Int. J. Biol. Macromol.* **2016**, *84*, 340–348. [CrossRef] [PubMed]
5. Amirian, J.; Zeng, Y.; Shekh, M.I.; Sharma, G.; Stadler, F.J.; Song, J.; Du, B.; Zhu, Y. In-situ crosslinked hydrogel based on amidated pectin/oxidized chitosan as potential wound dressing for skin repairing. *Carbohydr. Polym.* **2021**, *251*, 117005. [CrossRef]
6. Sharma, G.; Thakur, B.; Naushad, M.; Kumar, A.; Stadler, F.J.; Alfadul, S.M.; Mola, G.T. Applications of nanocomposite hydrogels for biomedical engineering and environmental protection. *Environ. Chem. Lett.* **2018**, *16*, 113–146. [CrossRef]
7. Bakshi, P.S.; Selvakumar, D.; Kadirvelu, K.; Kumar, N.S. Chitosan as an environment friendly biomaterial—A review on recent modifications and applications. *Int. J. Biol. Macromol.* **2020**, *150*, 1072–1083. [CrossRef]

8. Negm, N.A.; Hefni, H.H.H.; Abd-Elaal, A.A.A.; Badr, E.A.; Abou Kana, M.T.H. Advancement on modification of chitosan biopolymer and its potential applications. *Int. J. Biol. Macromol.* **2020**, *152*, 681–702. [CrossRef]
9. Salzano de Luna, M.; Sirignano, M. Upcycling soot particles into chitosan-based aerogels for water purification from organic pollutants. *J. Hazard. Mater.* **2021**, *2*, 100019. [CrossRef]
10. Irastorza, A.; Zarandona, I.; Andonegi, M.; Guerrero, P.; de la Caba, K. The versatility of collagen and chitosan: From food to biomedical applications. *Food Hydrocoll.* **2021**, *116*, 106633. [CrossRef]
11. Ahmed, T.; Noman, M.; Luo, J.; Muhammad, S.; Shahid, M.; Ali, M.A.; Zhang, M.; Li, B. Bioengineered chitosan-magnesium nanocomposite: A novel agricultural antimicrobial agent against *Acidovorax oryzae* and *Rhizoctonia solani* for sustainable rice production. *Int. J. Biol. Macromol.* **2021**, *168*, 834–845. [CrossRef]
12. Xavier, L.O.; Sganzerla, W.G.; Rosa, G.B.; da Rosa, C.G.; Agostinetto, L.; Veeck, A.P.L.; Bretanha, L.C.; Micke, G.A.; Dalla Costa, M.; Bertoldi, F.C.; et al. Chitosan packaging functionalized with *Cinnamodendron dinisii* essential oil loaded zein: A proposal for meat conservation. *Int. J. Biol. Macromol.* **2021**, *169*, 183–193. [CrossRef] [PubMed]
13. Hamed, I.; Özogul, F.; Regenstein, J.M. Industrial applications of crustacean by-products (chitin, chitosan, and chitooligosaccharides): A review. *Trends Food Sci. Technol.* **2016**, *48*, 40–50. [CrossRef]
14. Morin-Crini, N.; Lichtfouse, E.; Torri, G.; Crini, G. Fundamentals and Applications of Chitosan. In *Sustainable Agriculture Reviews 35: Chitin and Chitosan: History, Fundamentals and Innovations*; Crini, G., Lichtfouse, E., Eds.; Springer International Publishing: Cham, Switzerland, 2019; pp. 49–123.
15. Tyliszczak, B.; Drabczyk, A.; Kudłacik-Kramarczyk, S.; Sobczak-Kupiec, A. Sustainable Production of Chitosan. In *Sustainable Production: Novel Trends in Energy, Environment and Material Systems*; Królczyk, G.M., Wzorek, M., Król, A., Kochan, O., Su, J., Kacprzyk, J., Eds.; Springer International Publishing: Cham, Switzerland, 2020; pp. 45–60.
16. Venter, J.P.; Kotze, A.F.; Auzely-Velty, R.; Rinaudo, M. Synthesis and evaluation of the mucoadhesivity of a CD-chitosan derivative. *Int. J. Pharm.* **2006**, *313*, 36–42. [CrossRef] [PubMed]
17. Dutta, P.K.; Dutta, J.; Tripathi, V.S.; Research, I. Chitin and chitosan: Chemistry, properties and applications. *J. Sci. Ind.* **2004**, *63*, 20–31.
18. Priyadarshi, R.; Rhim, J.-W. Chitosan-based biodegradable functional films for food packaging applications. *Innov. Food Sci. Emerg. Technol.* **2020**, *62*, 102346. [CrossRef]
19. Tabriz, A.; Ur Rehman Alvi, M.A.; Khan Niazi, M.B.; Batool, M.; Bhatti, M.F.; Khan, A.L.; Khan, A.U.; Jamil, T.; Ahmad, N.M. Quaternized trimethyl functionalized chitosan based antifungal membranes for drinking water treatment. *Carbohydr. Polym.* **2019**, *207*, 17–25. [CrossRef]
20. Abd El-Hack, M.E.; El-Saadony, M.T.; Shafi, M.E.; Zabermawi, N.M.; Arif, M.; Batiha, G.E.; Khafaga, A.F.; Abd El-Hakim, Y.M.; Al-Sagheer, A.A. Antimicrobial and antioxidant properties of chitosan and its derivatives and their applications: A review. *Int. J. Biol. Macromol.* **2020**, *164*, 2726–2744. [CrossRef]
21. Li, J.; Zhuang, S. Antibacterial activity of chitosan and its derivatives and their interaction mechanism with bacteria: Current state and perspectives. *Eur. Polym. J.* **2020**, *138*, 109984. [CrossRef]
22. Yin, M.; Wang, Y.; Zhang, Y.; Ren, X.; Qiu, Y.; Huang, T.S. Novel quaternarized N-halamine chitosan and polyvinyl alcohol nanofibrous membranes as hemostatic materials with excellent antibacterial properties. *Carbohydr. Polym.* **2020**, *232*, 115823. [CrossRef]
23. Pandit, A.; Indurkar, A.; Deshpande, C.; Jain, R.; Dandekar, P. A systematic review of physical techniques for chitosan degradation. *Carbohydr. Polym. Technol. Appl.* **2021**, *2*, 100033. [CrossRef]
24. Ghahremanzadeh, F.; Alihosseini, F.; Semnani, D. Investigation and comparison of new galactosylation methods on PCL/chitosan scaffolds for enhanced liver tissue engineering. *Int. J. Biol. Macromol.* **2021**, *174*, 278–288. [CrossRef] [PubMed]
25. Gritsch, L.; Lovell, C.; Goldmann, W.H.; Boccaccini, A.R. Fabrication and characterization of copper(II)-chitosan complexes as antibiotic-free antibacterial biomaterial. *Carbohydr. Polym.* **2018**, *179*, 370–378. [CrossRef] [PubMed]
26. Kurita, K. Chitin and chitosan: Functional biopolymers from marine crustaceans. *Mar. Biotechnol.* **2006**, *8*, 203–226. [CrossRef] [PubMed]
27. Hoang, H.T.; Jo, S.H.; Phan, Q.T.; Park, H.; Park, S.H.; Oh, C.W.; Lim, K.T. Dual pH-/thermo-responsive chitosan-based hydrogels prepared using "click" chemistry for colon-targeted drug delivery applications. *Carbohydr. Polym.* **2021**, *260*, 117812. [CrossRef] [PubMed]
28. Ahsan, S.M.; Thomas, M.; Reddy, K.K.; Sooraparaju, S.G.; Asthana, A.; Bhatnagar, I. Chitosan as biomaterial in drug delivery and tissue engineering. *Int. J. Biol. Macromol.* **2018**, *110*, 97–109. [CrossRef]
29. Khan, M.A.; Mujahid, M. A review on recent advances in chitosan based composite for hemostatic dressings. *Int. J. Biol. Macromol.* **2019**, *124*, 138–147. [CrossRef]
30. Du, X.; Wu, L.; Yan, H.; Jiang, Z.; Li, S.; Li, W.; Bai, Y.; Wang, H.; Cheng, Z.; Kong, D.; et al. Microchannelled alkylated chitosan sponge to treat noncompressible hemorrhages and facilitate wound healing. *Nat. Commun.* **2021**, *12*, 4733. [CrossRef]
31. Dhivya, S.; Saravanan, S.; Sastry, T.P.; Selvamurugan, N. Nanohydroxyapatite-reinforced chitosan composite hydrogel for bone tissue repair in vitro and in vivo. *J. Nanobiotechnol.* **2015**, *13*, 40. [CrossRef]
32. Torr, K.M.; Chittenden, C.; Franich, R.A.; Kreber, B. Advances in understanding bioactivity of chitosan and chitosan oligomers against selected wood-inhabiting fungi. *Holzforschung* **2005**, *59*, 559–567. [CrossRef]

33. Pham, D.C.; Nguyen, T.H.; Ngoc, U.T.P.; Le, N.T.T.; Tran, T.V.; Nguyen, D.H. Preparation, Characterization and Antifungal Properties of Chitosan-Silver Nanoparticles Synergize Fungicide Against Pyricularia oryzae. *J. Nanosci. Nanotechnol.* **2018**, *18*, 5299–5305. [CrossRef]
34. Hong, H.-M.; Sim, G.-Y.; Park, S.-M.; Lee, E.-J.; Kim, D.-Y. Ameliorative Effect of Chitosan Complex on Miniature Pig Sperm Cryopreservation. *J. Emb. Trans.* **2018**, *33*, 337–342. [CrossRef]
35. Ahn, S.I.; Cho, S.; Choi, N.J. Effectiveness of Chitosan as a Dietary Supplement in Lowering Cholesterol in Murine Models: A Meta-Analysis. *Mar Drugs* **2021**, *19*, 26. [CrossRef] [PubMed]
36. Lutjohann, D.; Marinova, M.; Wolter, K.; Willinek, W.; Bitterlich, N.; Coenen, M.; Coch, C.; Stellaard, F. Influence of Chitosan Treatment on Surrogate Serum Markers of Cholesterol Metabolism in Obese Subjects. *Nutrients* **2018**, *10*, 72. [CrossRef] [PubMed]
37. Moramkar, N.; Bhatt, P. Insight into chitosan derived nanotherapeutics for anticancer drug delivery and imaging. *Eur. Polym. J.* **2021**, *154*, 110540. [CrossRef]
38. Hadi, J.M.; Aziz, S.B.; Nofal, M.M.; Hussen, S.A.; Hamsan, M.H.; Brza, M.A.; Abdulwahid, R.T.; Kadir, M.F.Z.; Woo, H.J. Electrical, Dielectric Property and Electrochemical Performances of Plasticized Silver Ion-Conducting Chitosan-Based Polymer Nanocomposites. *Membranes* **2020**, *10*, 151. [CrossRef]
39. Vorobiov, V.K.; Smirnov, M.A.; Bobrova, N.V.; Sokolova, M.P. Chitosan-supported deep eutectic solvent as bio-based electrolyte for flexible supercapacitor. *Mater. Lett.* **2021**, *283*, 128889. [CrossRef]
40. Desbrières, J.; Guibal, E. Chitosan for wastewater treatment. *Polym. Int.* **2018**, *67*, 7–14. [CrossRef]
41. Dodero, A.; Brunengo, E.; Alloisio, M.; Sionkowska, A.; Vicini, S.; Castellano, M. Chitosan-based electrospun membranes: Effects of solution viscosity, coagulant and crosslinker. *Carbohydr. Polym.* **2020**, *235*, 115976. [CrossRef]
42. Ferreira, L.M.B.; Dos Santos, A.M.; Boni, F.I.; Dos Santos, K.C.; Robusti, L.M.G.; de Souza, M.P.C.; Ferreira, N.N.; Carvalho, S.G.; Cardoso, V.M.O.; Chorilli, M.; et al. Design of chitosan-based particle systems: A review of the physicochemical foundations for tailored properties. *Carbohydr. Polym.* **2020**, *250*, 116968. [CrossRef]
43. Kordjazi, S.; Kamyab, K.; Hemmatinejad, N. Super-hydrophilic/oleophobic chitosan/acrylamide hydrogel: An efficient water/oil separation filter. *Adv. Compos. Hybrid Mater.* **2020**, *3*, 167–176. [CrossRef]
44. Zhou, G.; Wang, K.P.; Liu, H.W.; Wang, L.; Xiao, X.F.; Dou, D.D.; Fan, Y.B. Three-dimensional polylactic acid@graphene oxide/chitosan sponge bionic filter: Highly efficient adsorption of crystal violet dye. *Int. J. Biol. Macromol.* **2018**, *113*, 792–803. [CrossRef] [PubMed]
45. Hui, M.; Shengyan, P.; Yaqi, H.; Rongxin, Z.; Anatoly, Z.; Wei, C. A highly efficient magnetic chitosan "fluid" adsorbent with a high capacity and fast adsorption kinetics for dyeing wastewater purification. *Chem. Eng. J.* **2018**, *345*, 556–565. [CrossRef]
46. Urrutia, P.; Bernal, C.; Wilson, L.; Illanes, A. Use of chitosan heterofunctionality for enzyme immobilization: Beta-galactosidase immobilization for galacto-oligosaccharide synthesis. *Int. J. Biol. Macromol.* **2018**, *116*, 182–193. [CrossRef] [PubMed]
47. Pal, P.; Pal, A.; Nakashima, K.; Yadav, B.K. Applications of chitosan in environmental remediation: A review. *Chemosphere* **2021**, *266*, 128934. [CrossRef]
48. Mohammadzadeh Pakdel, P.; Peighambardoust, S.J. Review on recent progress in chitosan-based hydrogels for wastewater treatment application. *Carbohydr. Polym.* **2018**, *201*, 264–279. [CrossRef]
49. Rinaudo, M. Chitin and chitosan: Properties and applications. *Prog. Polym. Sci.* **2006**, *31*, 603–632. [CrossRef]
50. Brigham, C. Chitin and Chitosan: Sustainable, Medically Relevant Biomaterials. *Int. J. Biotech. Well. Indus.* **2017**, *6*, 41–47. [CrossRef]
51. Lehnert, R.J.; Kandelbauer, A. Comments on "Solubility parameter of chitin and chitosan" Carbohydrate Polymers 36 (1998) 121–127. *Carbohydr. Polym.* **2017**, *175*, 601–602. [CrossRef]
52. Ravindra, R.; Krovvidi, K.R.; Khan, A.A. Solubility parameter of chitin and chitosan. *Carbohydr. Polym.* **1998**, *36*, 121–127. [CrossRef]
53. Pardo-Castaño, C.; Bolaños, G. Solubility of chitosan in aqueous acetic acid and pressurized carbon dioxide-water: Experimental equilibrium and solubilization kinetics. *J. Supercrit. Fluids.* **2019**, *151*, 63–74. [CrossRef]
54. Cunha, R.A.; Soares, T.A.; Rusu, V.H.; Pontes, F.J.; Franca, E.F.; Lins, R.D. The Molecular Structure and Conformational Dynamics of Chitosan Polymers: An Integrated Perspective from Experiments and Computational Simulations. In *The Complex World of Polysaccharides*; BoD–Books on Demand: Norderstedt, Germany, 2012.
55. Uragami, T.T. *Material Science of Chitin and Chitosan*; Kodansha: Tokyo, Japan, 2006.
56. Liu, X.; Wu, Y.; Zhao, X.; Wang, Z. Fabrication and applications of bioactive chitosan-based organic-inorganic hybrid materials: A review. *Carbohydr. Polym.* **2021**, *267*, 118179. [CrossRef] [PubMed]
57. Bi, J.; Huang, X.; Wang, J.; Wang, T.; Wu, H.; Yang, J.; Lu, H.; Hao, H. Oil-phase cyclic magnetic adsorption to synthesize Fe_3O_4@C@TiO_2-nanotube composites for simultaneous removal of Pb(II) and Rhodamine B. *Chem. Eng. J.* **2019**, *366*, 50–61. [CrossRef]
58. Yan, Y.; Dong, X.; Sun, X.; Sun, X.; Li, J.; Shen, J.; Han, W.; Liu, X.; Wang, L. Conversion of waste FGD gypsum into hydroxyapatite for removal of Pb^{2+} and Cd^{2+} from wastewater. *J. Colloid Interface Sci.* **2014**, *429*, 68–76. [CrossRef] [PubMed]
59. Shahrajabian, M.H.; Chaski, C.; Polyzos, N.; Tzortzakis, N.; Petropoulos, S.A. Sustainable Agriculture Systems in Vegetable Production Using Chitin and Chitosan as Plant Biostimulants. *Biomolecules* **2021**, *11*, 819. [CrossRef]
60. Divya, K.; Jisha, M.S. Chitosan nanoparticles preparation and applications. *Environ. Chem. Lett.* **2018**, *16*, 101–112. [CrossRef]

61. Vidanarachchi, J.K.; Kim, S.K. *Chitin, Chitosan, Oligosaccharides and Their Derivatives. Biological Activities and Applications*; Kim, S.-K., Ed.; CRC Press: Boca Raton, FL, USA, 2010; p. 666.
62. Gutiérrez, T.J. Chapter 8: Chitosan Applications for the Food Industry. In *Chitosan: Derivatives, Composites and Applications*; Ahmed, S., Ed.; Wiley Online Library: Hoboken, NJ, USA, 2017.
63. Song, E.H.; Shang, J.; Ratner, D.M. 9.08-Polysaccharides. In *Polymer Science: A Comprehensive Reference*; Matyjaszewski, K., Möller, M., Eds.; Elsevier: Amsterdam, The Netherlands, 2012; pp. 137–155.
64. Negoi, A.-E.; Cristea, C.-D.; Deşliu-Avram, M.; Trică, B.; Constantinescu-Aruxandei, D.; Oancea, F. Extraction of Fungal Chitin Using Natural Deep Eutectic Solvents. *Proceedings* **2019**, *29*, 91.
65. Hongkulsup, C.; Khutoryanskiy, V.V.; Niranjan, K. Enzyme assisted extraction of chitin from shrimp shells (*Litopenaeus vannamei*). *J. Chem. Technol. Biotechnol.* **2016**, *91*, 1250–1256. [CrossRef]
66. Sadighara, P.; Moghadam, H.T.; Eskandari, S.; Salehi, A. Optimization of extraction of chitosan and carotenoids from shrimp waste. *J. Fish Aquat. Sci.* **2015**, *2*, 36–38.
67. Peniston, Q.P.; Johnson, E.L. Process for the Manufacture of Chitosan. U.S. Patent No. 4,195,175, 1980.
68. Kurita, K.; Kaji, Y.; Mori, T.; Nishiyama, Y. Enzymatic degradation of β-chitin: Susceptibility and the influence of deacetylation. *Carbohydr. Polym.* **2000**, *42*, 19–21. [CrossRef]
69. Tan, T.S.; Chin, H.Y.; Tsai, M.L.; Liu, C.L. Structural alterations, pore generation, and deacetylation of alpha- and beta-chitin submitted to steam e xplosion. *Carbohydr. Polym.* **2015**, *122*, 321–328. [CrossRef]
70. Anwar, M.; Anggraeni, A.S.; Amin, M.H.A. Comparison of green method for chitin deacetylation. *AIP Conf. Proc.* **2017**, *1823*, 020071.
71. Campana-Filho, S.P.; Signini, R.; Cardoso, M.B. Effects of sonication on the reactivity of chitin toward its heterogeneous deacetylation. *Int. J. Polym. Mater. Polym. Biomater.* **2002**, *51*, 695–700. [CrossRef]
72. Younes, I.; Ghorbel-Bellaaj, O.; Nasri, R.; Chaabouni, M.; Rinaudo, M.; Nasri, M. Chitin and chitosan preparation from shrimp shells using optimized enzymatic deproteinization. *Process Biochem.* **2012**, *47*, 2032–2039. [CrossRef]
73. Tasar, O.C.; Erdal, S.; Taskin, M. Chitosan production by psychrotolerant Rhizopus oryzae in non-sterile open fermentation conditions. *Int. J. Biol. Macromol.* **2016**, *89*, 428–433. [CrossRef] [PubMed]
74. Zhao, L.; Xia, W. Stainless steel membrane UF coupled with NF process for the recovery of sodium hydroxide from alkaline wastewater in chitin processing. *Desalination* **2009**, *249*, 774–780. [CrossRef]
75. Zhao, L.; Xia, W.; Zhao, H. Cost model for chitin production alkali wastewater recovery by couple-membrane filtration. *Desalin. Water Treat.* **2012**, *28*, 202–210. [CrossRef]
76. Jahan, M.S.; Hossain, M.M.; Roy, S.K.; Asaduzzaman, M.; Masum, S.M.; Nessa, F. A Process for the Preparation of Chitin and Chitosan from Prawn Shell Waste. *Bangladesh. J. Sci. Ind. Res.* **1970**, *45*, 323–330.
77. Muxika, A.; Etxabide, A.; Uranga, J.; Guerrero, P.; de la Caba, K. Chitosan as a bioactive polymer: Processing, properties and applications. *Int. J. Biol. Macromol.* **2017**, *105*, 1358–1368. [CrossRef]
78. Deng, J.; Zhu, E.-Q.; Xu, G.-F.; Naik, N.; Murugadoss, V.; Ma, M.-G.; Guo, Z.; Shi, Z.-J. Overview of renewable polysaccharide-based composites for biodegradable food packaging applications. *Green Chem.* **2022**, *24*, 480–492. [CrossRef]
79. Sharma, G.; Khosla, A.; Kumar, A.; Kaushal, N.; Sharma, S.; Naushad, M.; Vo, D.-V.N.; Iqbal, J.; Stadler, F.J. A comprehensive review on the removal of noxious pollutants using carrageenan based advanced adsorbents. *Chemosphere* **2022**, *289*, 133100. [CrossRef]
80. Sharma, G.; Kumar, A.; Ghfar, A.A.; García-Peñas, A.; Naushad, M.; Stadler, F.J. Fabrication and Characterization of Xanthan Gum-cl-poly(acrylamide-co-alginic acid) Hydrogel for Adsorption of Cadmium Ions from Aqueous Medium. *Gels* **2022**, *8*, 23. [CrossRef] [PubMed]
81. Sharma, G.; Kumar, A.; Chauhan, C.; Okram, A.; Sharma, S.; Pathania, D.; Kalia, S. Pectin-crosslinked-guar gum/SPION nanocomposite hydrogel for adsorption of m-cresol and o-chlorophenol. *Sustain. Chem. Pharm.* **2017**, *6*, 96–106. [CrossRef]
82. Elanchezhiyan, S.S.; Preethi, J.; Rathinam, K.; Njaramba, L.K.; Park, C.M. Synthesis of magnetic chitosan biopolymeric spheres and their adsorption performances for PFOA and PFOS from aqueous environment. *Carbohydr. Polym.* **2021**, *267*, 118165. [CrossRef]
83. Sharma, G.; Kumar, A.; Naushad, M.; Thakur, B.; Vo, D.-V.N.; Gao, B.; Al-Kahtani, A.A.; Stadler, F.J. Adsorptional-photocatalytic removal of fast sulphon black dye by using chitin-cl-poly(itaconic acid-co-acrylamide)/zirconium tungstate nanocomposite hydrogel. *J. Hazard. Mater.* **2021**, *416*, 125714. [CrossRef] [PubMed]
84. Leceta, I.; Molinaro, S.; Guerrero, P.; Kerry, J.P.; de la Caba, K. Quality attributes of map packaged ready-to-eat baby carrots by using chitosan-based coatings. *Postharvest Biol. Technol.* **2015**, *100*, 142–150. [CrossRef]
85. Liu, S.; Gao, J.; Zhang, L.; Yang, Y.; Liu, X. Diethylenetriaminepentaacetic acid–thiourea-modified magnetic chitosan for adsorption of hexavalent chromium from aqueous solutions. *Carbohydr. Polym.* **2021**, *274*, 118555. [CrossRef]
86. Kumar, S.; Mukherjee, A.; Dutta, J. Chitosan based nanocomposite films and coatings: Emerging antimicrobial food packaging alternatives. *Trends Food Sci. Technol.* **2020**, *97*, 196–209. [CrossRef]
87. Elmehbad, N.Y.; Mohamed, N.A.; Abd El-Ghany, N.A. Evaluation of the antimicrobial and anti-biofilm activity of novel salicylhydrazido chitosan derivatives impregnated with titanium dioxide nanoparticles. *Int. J. Biol. Macromol.* **2022**, *205*, 719–730. [CrossRef]
88. Rodrigues, P.R.; Junior, L.M.; de Souza, W.F.C.; Sato, H.H.; Alves, R.M.V.; Vieira, R.P. O-ATRP synthesized poly(β-pinene) blended with chitosan for antimicrobial and antioxidant bio-based films production. *Int. J. Biol. Macromol.* **2021**, *193*, 425–432. [CrossRef]

89. Nadira, P.P.; Mujeeb, V.M.A.; Rahman, P.M.; Muraleedharan, K. Effects of cashew leaf extract on physicochemical, antioxidant, and antimicrobial properties of N, O–Carboxymethyl chitosan films. *Carbohydr. Polym. Technol. Appl.* **2022**, *3*, 100191. [CrossRef]
90. Elsabee, M.Z.; Abdou, E.S. Chitosan based edible films and coatings: A review. *Mater Sci. Eng. C Mater. Biol. Appl.* **2013**, *33*, 1819–1841. [CrossRef] [PubMed]
91. Haghighi, H.; Licciardello, F.; Fava, P.; Siesler, H.W.; Pulvirenti, A. Recent advances on chitosan-based films for sustainable food packaging applications. *Food Packag. Shelf Life* **2020**, *26*, 100551–100567. [CrossRef]
92. Panda, P.K.; Dash, P.; Yang, J.-M.; Chang, Y.-H. Development of chitosan, graphene oxide, and cerium oxide composite blended films: Structural, physical, and functional properties. *Cellulose* **2022**, *29*, 2399–2411. [CrossRef]
93. Kim, D.S.; Dhand, V.; Rhee, K.Y.; Park, S.-J. Study on the Effect of Silanization and Improvement in the Tensile Behavior of Graphene-Chitosan-Composite. *Polymers* **2015**, *7*, 527–551. [CrossRef]
94. Sharma, G.; Naushad, M.; Kumar, A.; Kumar, A.; Ahamad, T.; Stadler, F.J. Facile fabrication of chitosan-cl-poly(AA)/$ZrPO_4$ nanocomposite for remediation of rhodamine B and antimicrobial activity. *J. King Saud Univ. Sci.* **2020**, *32*, 1359–1365. [CrossRef]
95. You, J.; Liu, C.; Feng, X.; Lu, B.; Xia, L.; Zhuang, X. In situ synthesis of ZnS nanoparticles onto cellulose/chitosan sponge for adsorption–photocatalytic removal of Congo red. *Carbohydr. Polym.* **2022**, *288*, 119332. [CrossRef]
96. Andrade-Del Olmo, J.; Pérez-Álvarez, L.; Hernáez, E.; Ruiz-Rubio, L.; Vilas-Vilela, J.L. Antibacterial multilayer of chitosan and (2-carboxyethyl)- β-cyclodextrin onto polylactic acid (PLLA). *Food Hydrocoll.* **2019**, *88*, 228–236. [CrossRef]
97. Zhang, M.; Zheng, Y.; Jin, Y.; Wang, D.; Wang, G.; Zhang, X.; Li, Y.; Lee, S. Ag@MOF-loaded p-coumaric acid modified chitosan/chitosan nanoparticle and polyvinyl alcohol/starch bilayer films for food packing applications. *Int. J. Biol. Macromol.* **2022**, *202*, 80–90. [CrossRef]
98. Zhang, H.; He, P.; Kang, H.; Li, X. Antioxidant and antimicrobial effects of edible coating based on chitosan and bamboo vinegar in ready to cook pork chops. *LWT* **2018**, *93*, 470–476. [CrossRef]
99. Balti, R.; Mansour, M.B.; Sayari, N.; Yacoubi, L.; Rabaoui, L.; Brodu, N.; Massé, A. Development and characterization of bioactive edible films from spider crab (*Maja crispata*) chitosan incorporated with Spirulina extract. *Int. J. Biol. Macromol.* **2017**, *105*, 1464–1472. [CrossRef]
100. Ghaderi-Ghahfarokhi, M.; Barzegar, M.; Sahari, M.A.; Ahmadi Gavlighi, H.; Gardini, F. Chitosan-cinnamon essential oil nano-formulation: Application as a novel additive for controlled release and shelf life extension of beef patties. *Int. J. Biol. Macromol.* **2017**, *102*, 19–28. [CrossRef] [PubMed]
101. He, L.; Zou, L.; Yang, Q.; Xia, J.; Zhou, K.; Zhu, Y.; Han, X.; Pu, B.; Hu, B.; Deng, W.; et al. Antimicrobial Activities of Nisin, Tea Polyphenols, and Chitosan and their Combinations in Chilled Mutton. *J. Food Sci.* **2016**, *81*, M1466–M1471. [CrossRef] [PubMed]
102. Khanjari, A.; Karabagias, I.K.; Kontominas, M.G. Combined effect of N,O-carboxymethyl chitosan and oregano essential oil to extend shelf life and control Listeria monocytogenes in raw chicken meat fillets. *LWT Food Sci. Technol.* **2013**, *53*, 94–99. [CrossRef]
103. Bautista-Baños, S.; Hernández-López, M.; Bosquez-Molina, E.; Wilson, C.L. Effects of chitosan and plant extracts on growth of Colletotrichum gloeosporioides, anthracnose levels and quality of papaya fruit. *Crop Prot.* **2003**, *22*, 1087–1092. [CrossRef]
104. Chamanara, V.; Shabanpour, B.; Gorgin, S.; Khomeiri, M. An investigation on characteristics of rainbow trout coated using chitosan assisted with thyme essential oil. *Int. J. Biol. Macromol.* **2012**, *50*, 540–544. [CrossRef] [PubMed]
105. Liu, Y.; Wang, D.; Sun, Z.; Liu, F.; Du, L.; Wang, D. Preparation and characterization of gelatin/chitosan/3-phenylacetic acid food-packaging nanofiber antibacterial films by electrospinning. *Int. J. Biol. Macromol.* **2021**, *169*, 161–170. [CrossRef]
106. Wang, H.; Ding, F.; Ma, L.; Zhang, Y. Edible films from chitosan-gelatin: Physical properties and food packaging application. *Food Biosci.* **2021**, *40*, 100871. [CrossRef]
107. Fasciglione, G.; Goñi, M.G.; Yommi, A.K.; Perez-Bravo, J.J.; Ortueta, R.; Scampini, A.; Buffa, L.; Andreu, A.B.; Creus, C.M. Revaluation of waste from fishing industry through generation of chitosan coatings to improve quality and extend shelf-life of minimally processed lettuce. *Postharvest Biol. Technol.* **2020**, *170*, 111310. [CrossRef]
108. Panda, P.K.; Yang, J.-M.; Chang, Y.-H. Preparation and characterization of ferulic acid-modified water soluble chitosan and poly (γ-glutamic acid) polyelectrolyte films through layer-by-layer assembly towards protein adsorption. *Int. J. Biol. Macromol.* **2021**, *171*, 457–464. [CrossRef]
109. Ngo, T.M.P.; Nguyen, T.H.; Dang, T.M.Q.; Tran, T.X.; Rachtanapun, P. Characteristics and Antimicrobial Properties of Active Edible Films Based on Pectin and Nanochitosan. *Int. J. Mol. Sci.* **2020**, *21*, 2224. [CrossRef]
110. Baron, R.D.; Perez, L.L.; Salcedo, J.M.; Cordoba, L.P.; Sobral, P.J. Production and characterization of films based on blends of chitosan from blue crab (*Callinectes sapidus*) waste and pectin from Orange (Citrus sinensis Osbeck) peel. *Int. J. Biol. Macromol.* **2017**, *98*, 676–683. [CrossRef] [PubMed]
111. Valizadeh, S.; Naseri, M.; Babaei, S.; Hosseini, S.M.H.; Imani, A. Development of bioactive composite films from chitosan and carboxymethyl cellulose using glutaraldehyde, cinnamon essential oil and oleic acid. *Int. J. Biol. Macromol.* **2019**, *134*, 604–612. [CrossRef] [PubMed]
112. Youssef, A.M.; El-Sayed, S.M.; El-Sayed, H.S.; Salama, H.H.; Dufresne, A. Enhancement of Egyptian soft white cheese shelf life using a novel chitosan/carboxymethyl cellulose/zinc oxide bionanocomposite film. *Carbohydr. Polym.* **2016**, *151*, 9–19. [CrossRef]
113. Hu, D.; Wang, H.; Wang, L. Physical properties and antibacterial activity of quaternized chitosan/carboxymethyl cellulose blend films. *LWT Food Sci. Technol.* **2016**, *65*, 398–405. [CrossRef]
114. Xu, T.; Gao, C.; Feng, X.; Yang, Y.; Shen, X.; Tang, X. Structure, physical and antioxidant properties of chitosan-gum arabic edible films incorporated with cinnamon essential oil. *Int. J. Biol. Macromol.* **2019**, *134*, 230–236. [CrossRef] [PubMed]

115. Xu, T.; Gao, C.; Feng, X.; Huang, M.; Yang, Y.; Shen, X.; Tang, X. Cinnamon and clove essential oils to improve physical, thermal and antimicrobial properties of chitosan-gum arabic polyelectrolyte complexed films. *Carbohydr. Polym.* **2019**, *217*, 116–125. [CrossRef] [PubMed]
116. Luchese, C.L.; Pavoni, J.M.F.; Dos Santos, N.Z.; Quines, L.K.; Pollo, L.D.; Spada, J.C.; Tessaro, I.C. Effect of chitosan addition on the properties of films prepared with corn and cassava starches. *J. Food Sci. Technol.* **2018**, *55*, 2963–2973. [CrossRef]
117. Ren, L.; Yan, X.; Zhou, J.; Tong, J.; Su, X. Influence of chitosan concentration on mechanical and barrier properties of corn starch/chitosan films. *Int. J. Biol. Macromol.* **2017**, *105*, 1636–1643. [CrossRef]
118. Lozano-Navarro, J.I.; Diaz-Zavala, N.P.; Velasco-Santos, C.; Martinez-Hernandez, A.L.; Tijerina-Ramos, B.I.; Garcia-Hernandez, M.; Rivera-Armenta, J.L.; Paramo-Garcia, U.; Reyes-de la Torre, A.I. Antimicrobial, Optical and Mechanical Properties of Chitosan-Starch Films with Natural Extracts. *Int. J. Mol. Sci.* **2017**, *18*, 997. [CrossRef]
119. Liu, Y.; Yuan, Y.; Duan, S.; Li, C.; Hu, B.; Liu, A.; Wu, D.; Cui, H.; Lin, L.; He, J.; et al. Preparation and characterization of chitosan films with three kinds of molecular weight for food packaging. *Int. J. Biol. Macromol.* **2020**, *155*, 249–259. [CrossRef]
120. Zhao, J.; Wei, F.; Xu, W.; Han, X. Enhanced antibacterial performance of gelatin/chitosan film containing capsaicin loaded MOFs for food packaging. *Appl. Surf. Sci.* **2020**, *510*, 145418. [CrossRef]
121. Rambabu, K.; Bharath, G.; Banat, F.; Show, P.L.; Cocoletzi, H.H. Mango leaf extract incorporated chitosan antioxidant film for active food packaging. *Int. J. Biol. Macromol.* **2019**, *126*, 1234–1243.
122. Kumar, S.; Shukla, A.; Baul, P.P.; Mitra, A.; Halder, D. Biodegradable hybrid nanocomposites of chitosan/gelatin and silver nanoparticles for active food packaging applications. *Food Packag. Shelf Life* **2018**, *16*, 178–184. [CrossRef]
123. Zhang, L.; Huang, C.; Zhao, H. Application of Pullulan and Chitosan Multilayer Coatings in Fresh Papayas. *Coatings* **2019**, *9*, 745. [CrossRef]
124. Kaewklin, P.; Siripatrawan, U.; Suwanagul, A.; Lee, Y.S. Active packaging from chitosan-titanium dioxide nanocomposite film for prolonging storage life of tomato fruit. *Int. J. Biol. Macromol.* **2018**, *112*, 523–529. [CrossRef]
125. Gao, Q.; Lei, M.; Zhou, K.; Liu, X.; Wang, S.; Li, H. Preparation of a microfibrillated cellulose/chitosan/polypyrrole film for Active Food Packaging. *Prog. Org. Coat.* **2020**, *149*, 105907. [CrossRef]
126. Lu, S.; Tao, J.; Liu, X.; Wen, Z. Baicalin-liposomes loaded polyvinyl alcohol-chitosan electrospinning nanofibrous films: Characterization, antibacterial properties and preservation effects on mushrooms. *Food Chem.* **2022**, *371*, 131372. [CrossRef]
127. Azaza, Y.B.; Hamdi, M.; Charmette, C.; Jridi, M.; Li, S.; Nasri, M.; Nasri, R. Development and characterization of active packaging films based on chitosan and sardinella protein isolate: Effects on the quality and the shelf life of shrimps. *Food Packag. Shelf Life* **2022**, *31*, 100796. [CrossRef]
128. Lin, L.; Xue, L.; Duraiarasan, S.; Haiying, C. Preparation of ε-polylysine/chitosan nanofibers for food packaging against Salmonella on chicken. *Food Packag. Shelf Life* **2018**, *17*, 134–141. [CrossRef]
129. Priyadarshi, R.; Kumar, B.; Deeba, F.; Kulshreshtha, A.; Negi, Y.S. Chitosan films incorporated with Apricot (Prunus armeniaca) kernel essential oil as active food packaging material. *Food Hydrocoll.* **2018**, *85*, 158–166. [CrossRef]
130. Cui, H.; Surendhiran, D.; Li, C.; Lin, L. Biodegradable zein active film containing chitosan nanoparticle encapsulated with pomegranate peel extract for food packaging. *Food Packag. Shelf Life* **2020**, *24*, 100511. [CrossRef]
131. Indumathi, M.P.; Rajarajeswari, G.R. Mahua oil-based polyurethane/chitosan/nano ZnO composite films for biodegradable food packaging applications. *Int. J. Biol. Macromol.* **2019**, *124*, 163–174.
132. Liu, X.; Xue, F.; Li, C.; Adhikari, B. Physicochemical properties of films produced using nanoemulsions stabilized by carboxymethyl chitosan-peptide conjugates and application in blueberry preservation. *Int. J. Biol. Macromol.* **2022**, *202*, 26–36. [CrossRef] [PubMed]
133. Vilvert, J.C.; de Freitas, S.T.; Ferreira, M.A.R.; Leite, R.H.d.L.; dos Santos, F.K.G.; Costa, C.d.S.R.; Aroucha, E.M.M. Chitosan and graphene oxide-based biodegradable bags: An eco-friendly and effective packaging alternative to maintain postharvest quality of 'Palmer' mango. *LWT* **2022**, *154*, 112741. [CrossRef]
134. Yang, K.; Dang, H.; Liu, L.; Hu, X.; Li, X.; Ma, Z.; Wang, X.; Ren, T. Effect of syringic acid incorporation on the physical, mechanical, structural and antibacterial properties of chitosan film for quail eggs preservation. *Int. J. Biol. Macromol.* **2019**, *141*, 876–884. [CrossRef]
135. Yao, X.; Hu, H.; Qin, Y.; Liu, J. Development of antioxidant, antimicrobial and ammonia-sensitive films based on quaternary ammonium chitosan, polyvinyl alcohol and betalains-rich cactus pears (*Opuntia ficus-indica*) extract. *Food Hydrocoll.* **2020**, *106*, 105896. [CrossRef]
136. Duran, A.; Kahve, H.I. The effect of chitosan coating and vacuum packaging on the microbiological and chemical properties of beef. *Meat Sci.* **2020**, *162*, 107961. [CrossRef]
137. Liu, Z.; Du, M.; Liu, H.; Zhang, K.; Xu, X.; Liu, K.; Tu, J.; Liu, Q. Chitosan films incorporating litchi peel extract and titanium dioxide nanoparticles and their application as coatings on watercored apples. *Prog. Org. Coat.* **2021**, *151*, 106103. [CrossRef]
138. Fathima, P.E.; Panda, S.K.; Ashraf, P.M.; Varghese, T.O.; Bindu, J. Polylactic acid/chitosan films for packaging of Indian white prawn (*Fenneropenaeus indicus*). *Int. J. Biol. Macromol.* **2018**, *117*, 1002–1010. [CrossRef]
139. Nagarajan, M.; Rajasekaran, B.; Benjakul, S.; Venkatachalam, K. Influence of chitosan-gelatin edible coating incorporated with longkong pericarp extract on refrigerated black tiger Shrimp (*Penaeus monodon*). *Curr. Res. Food Sci.* **2021**, *4*, 345–353. [CrossRef]

140. Merlo, T.C.; Contreras-Castillo, C.J.; Saldaña, E.; Barancelli, G.V.; Dargelio, M.D.B.; Yoshida, C.M.P.; Ribeiro Junior, E.E.; Massarioli, A.; Venturini, A.C. Incorporation of pink pepper residue extract into chitosan film combined with a modified atmosphere packaging: Effects on the shelf life of salmon fillets. *Food Res. Int.* **2019**, *125*, 108633. [CrossRef] [PubMed]
141. Priyadarshi, R.; Kumar, B.; Negi, Y.S. Chitosan film incorporated with citric acid and glycerol as an active packaging material for extension of green chilli shelf life. *Carbohydr. Polym.* **2018**, *195*, 329–338. [CrossRef] [PubMed]
142. Orzali, L.; Corsi, B.; Forni, C.; Riccioni, L. Chitosan in Agriculture: A New Challenge for Managing Plant Disease. In *Biological Activities and Application of Marine Polysaccharides*; BoD–Books on Demand: Norderstedt, Germany, 2017.
143. Kulikov, S.N.; Chirkov, S.N.; Il'ina, A.V.; Lopatin, S.A.; Varlamov, V.P. Effect of the molecular weight of chitosan on its antiviral activity in plants. *Prikl. Biokhim. Mikrobiol.* **2006**, *42*, 224–228. [CrossRef]
144. Hadwiger, L.A. Multiple effects of chitosan on plant systems: Solid science or hype. *Plant Sci.* **2013**, *208*, 42–49. [CrossRef] [PubMed]
145. Adamuchio-Oliveira, L.G.; Mazaro, S.M.; Mógor, G.; Sant'Anna-Santos, B.F.; Mógor, Á.F. Chitosan associated with chelated copper applied on tomatoes: Enzymatic and anatomical changes related to plant defense responses. *Sci. Hortic.* **2020**, *271*, 109431. [CrossRef]
146. Jogaiah, S.; Satapute, P.; De Britto, S.; Konappa, N.; Udayashankar, A.C. Exogenous priming of chitosan induces upregulation of phytohormones and resistance against cucumber powdery mildew disease is correlated with localized biosynthesis of defense enzymes. *Int. J. Biol. Macromol.* **2020**, *162*, 1825–1838. [CrossRef]
147. Alkahtani, M.D.F.; Attia, K.A.; Hafez, Y.M.; Khan, N.; Eid, A.M.; Ali, M.A.M.; Abdelaal, K.A.A. Chlorophyll Fluorescence Parameters and Antioxidant Defense System Can Display Salt Tolerance of Salt Acclimated Sweet Pepper Plants Treated with Chitosan and Plant Growth Promoting Rhizobacteria. *Agronomy* **2020**, *10*, 1180. [CrossRef]
148. Kahromi, S.; Khara, J. Chitosan stimulates secondary metabolite production and nutrient uptake in medicinal plant *Dracocephalum kotschyi*. *J. Sci. Food Agric.* **2021**, *101*, 3898–3907. [CrossRef]
149. Rasheed, R.; Ashraf, M.A.; Arshad, A.; Iqbal, M.; Hussain, I. Interactive effects of chitosan and cadmium on growth, secondary metabolism, oxidative defense, and element uptake in pea (*Pisum sativum* L.). *Arabian J. Geosci.* **2020**, *13*, 847. [CrossRef]
150. Kumari, S.; Singh, A.K.; Kumar, A.; Singh, K.P.; Bains, G. Evaluating the efficacy of chitosan and salicylic acid on photosynthetic pigments and antioxidant enzymes towards resistance of mango malformation. *Sci. Hortic.* **2021**, *285*, 110160. [CrossRef]
151. Picchi, V.; Gobbi, S.; Fattizzo, M.; Zefelippo, M.; Faoro, F. Chitosan Nanoparticles Loaded with N-Acetyl Cysteine to Mitigate Ozone and Other Possible Oxidative Stresses in Durum Wheat. *Plants* **2021**, *10*, 691. [CrossRef] [PubMed]
152. Li, Y.; Zhang, Q.; Ou, L.; Ji, D.; Liu, T.; Lan, R.; Li, X.; Jin, L. Response to the Cold Stress Signaling of the Tea Plant (*Camellia sinensis*) Elicited by Chitosan Oligosaccharide. *Agronomy* **2020**, *10*, 915. [CrossRef]
153. Chaudhari, A.K.; Singh, V.K.; Das, S.; Dubey, N.K. Fabrication, characterization, and bioactivity assessment of chitosan nanoemulsion containing allspice essential oil to mitigate *Aspergillus flavus* contamination and aflatoxin B1 production in maize. *Food Chem.* **2022**, *372*, 131221. [CrossRef] [PubMed]
154. Mondéjar-López, M.; Rubio-Moraga, A.; López-Jimenez, A.J.; García Martínez, J.C.; Ahrazem, O.; Gómez-Gómez, L.; Niza, E. Chitosan nanoparticles loaded with garlic essential oil: A new alternative to tebuconazole as seed dressing agent. *Carbohydr. Polym.* **2022**, *277*, 118815. [CrossRef]
155. Wu, P.; Xin, F.; Xu, H.; Chu, Y.; Du, Y.; Tian, H.; Zhu, B. Chitosan inhibits postharvest berry abscission of 'Kyoho' table grapes by affecting the structure of abscission zone, cell wall degrading enzymes and SO_2 permeation. *Postharvest Biol. Technol.* **2021**, *176*, 111507. [CrossRef]
156. Li, Z.; Xue, S.; Xu, X.; Wang, B.; Zheng, X.; Li, B.; Xie, P.; Bi, Y.; Prusky, D. Preharvest multiple sprays with chitosan accelerate the deposition of suberin poly phenolic at wound sites of harvested muskmelons. *Postharvest Biol. Technol.* **2021**, *179*, 111565. [CrossRef]
157. Jiang, Y.; Yin, H.; Zhou, X.; Wang, D.; Zhong, Y.; Xia, Q.; Deng, Y.; Zhao, Y. Antimicrobial, antioxidant and physical properties of chitosan film containing *Akebia trifoliata* (Thunb.) Koidz. peel extract/montmorillonite and its application. *Food Chem.* **2021**, *361*, 130111. [CrossRef]
158. Kumar, A.; Pratap Singh, P.; Prakash, B. Unravelling the antifungal and anti-aflatoxin B1 mechanism of chitosan nanocomposite incorporated with *Foeniculum vulgare* essential oil. *Carbohydr. Polym.* **2020**, *236*, 116050. [CrossRef]
159. Rajkumar, V.; Gunasekaran, C.; Paul, C.A.; Dharmaraj, J. Development of encapsulated peppermint essential oil in chitosan nanoparticles: Characterization and biological efficacy against stored-grain pest control. *Pestic. Biochem. Physiol.* **2020**, *170*, 104679. [CrossRef]
160. Kangama, A.; Zeng, D.; Tian, X.; Fang, J. Application of Chitosan Composite Flocculant in Tap Water Treatment. *J. Chem.* **2018**, *2018*, 2768474. [CrossRef]
161. Abhinaya, M.; Parthiban, R.; Kumar, P.S.; Vo, D.-V.N. A review on cleaner strategies for extraction of chitosan and its application in toxic pollutant removal. *Environ. Res.* **2021**, *196*, 110996. [CrossRef] [PubMed]
162. Sadiq, A.C.; Olasupo, A.; Ngah, W.S.W.; Rahim, N.Y.; Suah, F.B.M. A decade development in the application of chitosan-based materials for dye adsorption: A short review. *Int. J. Biol. Macromol.* **2021**, *191*, 1151–1163. [CrossRef] [PubMed]
163. Saheed, I.O.; Oh, W.D.; Suah, F.B.M. Chitosan modifications for adsorption of pollutants—A review. *J. Hazard. Mater.* **2021**, *408*, 124889. [CrossRef] [PubMed]

164. Li, Y.; Liang, Y.-Q.; Mao, X.-M.; Li, H. Efficient removal of Cu(II) from an aqueous solution using a novel chitosan assisted EDTA-intercalated hydrotalcite-like compound composite: Preparation, characterization, and adsorption mechanism. *Chem. Eng. J.* **2022**, *438*, 135531. [CrossRef]
165. Alsamman, M.T.; Sánchez, J. Recent advances on hydrogels based on chitosan and alginate for the adsorption of dyes and metal ions from water. *Arabian J. Chem.* **2021**, *14*, 103455. [CrossRef]
166. Abdolmaleki, A.Y.; Zilouei, H.; Khorasani, S.N.; Zargoosh, K. Adsorption of tetracycline from water using glutaraldehyde-crosslinked electrospun nanofibers of chitosan/poly(vinyl alcohol). *Water Sci. Technol.* **2018**, *77*, 1324–1335. [CrossRef]
167. Afzal, M.Z.; Sun, X.F.; Liu, J.; Song, C.; Wang, S.G.; Javed, A. Enhancement of ciprofloxacin sorption on chitosan/biochar hydrogel beads. *Sci. Total Environ.* **2018**, *639*, 560–569. [CrossRef]
168. Ahamad, T.; Chaudhary, A.A.; Naushad, M.; Alshehri, S.M. Fabrication of $MnFe_2O_4$ nanoparticles embedded chitosan-diphenylureaformaldehyde resin for the removal of tetracycline from aqueous solution. *Int. J. Biol. Macromol.* **2019**, *134*, 180–188. [CrossRef]
169. Ahamad, T.; Naushad, M.; Al-Shahrani, T.; Al-Hokbany, N.; Alshehri, S.M. Preparation of chitosan based magnetic nanocomposite for tetracycline adsorption: Kinetic and thermodynamic studies. *Int. J. Biol. Macromol.* **2020**, *147*, 258–267. [CrossRef]
170. Danalioglu, S.T.; Kerkez Kuyumcu, O.; Abdel Salam, M.; Bayazit, S.S. Chitosan grafted SiO_2-Fe_3O_4 nanoparticles for removal of antibiotics from water. *Environ. Sci. Pollut. Res.* **2018**, *25*, 36661–36670. [CrossRef]
171. Chen, J.; Ouyang, J.; Chen, W.; Zheng, Z.; Yang, Z.; Liu, Z.; Zhou, L. Fabrication and adsorption mechanism of chitosan/Zr-MOF (UiO-66) composite foams for efficient removal of ketoprofen from aqueous solution. *Chem. Eng. J.* **2022**, *431*, 134045. [CrossRef]
172. Tang, X.; Huang, Y.; He, Q.; Wang, Y.; Zheng, H.; Hu, Y. Adsorption of tetracycline antibiotics by nitrilotriacetic acid modified magnetic chitosan-based microspheres from aqueous solutions. *Environ. Technol. Innov.* **2021**, *24*, 101895. [CrossRef]
173. Rezaei, H.; Razavi, A.; Shahbazi, A. Removal of Congo red from aqueous solutions using nano-Chitosan. *Environ. Resour. Res.* **2017**, *5*, 25–34.
174. Ulu, A.; Alpaslan, M.; Gultek, A.; Ates, B. Eco-friendly chitosan/κ-carrageenan membranes reinforced with activated bentonite for adsorption of methylene blue. *Mater. Chem. Phys.* **2022**, *278*, 125611. [CrossRef]
175. Vieira, M.L.G.; Esquerdo, V.M.; Nobre, L.R.; Dotto, G.L.; Pinto, L.A.A. Glass beads coated with chitosan for the food azo dyes adsorption in a fixed bed column. *J. Ind. Eng. Chem.* **2014**, *20*, 3387–3393. [CrossRef]
176. Rathinam, K.; Singh, S.P.; Arnusch, C.J.; Kasher, R. An environmentally-friendly chitosan-lysozyme biocomposite for the effective removal of dyes and heavy metals from aqueous solutions. *Carbohydr. Polym.* **2018**, *199*, 506–515. [CrossRef]
177. Liu, X.; Zhang, Y.; Ju, H.; Yang, F.; Luo, X.; Zhang, L. Uptake of methylene blue on divinylbenzene cross-linked chitosan/maleic anhydride polymer by adsorption process. *Colloids Surf. A* **2021**, *629*, 127424. [CrossRef]
178. Kalidason, A.; Kuroiwa, T. Synthesis of chitosan–magnetite gel microparticles with improved stability and magnetic properties: A study on their adsorption, recoverability, and reusability in the removal of monovalent and multivalent azo dyes. *React. Funct. Polym.* **2022**, *173*, 105220. [CrossRef]
179. Abdul Mubarak, N.S.; Chuan, T.W.; Khor, H.P.; Jawad, A.H.; Wilson, L.D.; Sabar, S. Immobilized Fe-Loaded Chitosan Film for Methyl Orange Dye Removal: Competitive Ions, Reusability, and Mechanism. *J. Polym. Environ.* **2021**, *29*, 1050–1062. [CrossRef]
180. Tanhaei, B.; Ayati, A.; Iakovleva, E.; Sillanpää, M. Efficient carbon interlayed magnetic chitosan adsorbent for anionic dye removal: Synthesis, characterization and adsorption study. *Int. J. Biol. Macromol.* **2020**, *164*, 3621–3631. [CrossRef]
181. Taher, F.A.; Kamal, F.H.; Badawy, N.A.; Shrshr, A.E. Hierarchical magnetic/chitosan/graphene oxide 3D nanostructure as highly effective adsorbent. *Mater. Res. Bull.* **2018**, *97*, 361–368. [CrossRef]
182. Abdulhameed, A.S.; Mohammad, A.-T.; Jawad, A.H. Application of response surface methodology for enhanced synthesis of chitosan tripolyphosphate/TiO_2 nanocomposite and adsorption of reactive orange 16 dye. *J. Cleaner Prod.* **2019**, *232*, 43–56. [CrossRef]
183. Subramaniam, S.; Foo, K.Y.; Md Yusof, E.N.; Jawad, A.H.; Wilson, L.D.; Sabar, S. Hydrothermal synthesis of phosphorylated chitosan and its adsorption performance towards Acid Red 88 dye. *Int. J. Biol. Macromol.* **2021**, *193*, 1716–1726. [CrossRef]
184. Rostamian, M.; Hosseini, H.; Fakhri, V.; Talouki, P.Y.; Farahani, M.; Gharehtzpeh, A.J.; Goodarzi, V.; Su, C.-H. Introducing a bio sorbent for removal of methylene blue dye based on flexible poly(glycerol sebacate)/chitosan/graphene oxide ecofriendly nanocomposites. *Chemosphere* **2022**, *289*, 133219. [CrossRef] [PubMed]
185. Chen, M.; Bai, Y.; Liu, J.; Liu, Y.; Wang, Z.; Feng, X. Adsorption properties of magnetic sodium ferrosilicate/carboxymethyl chitosan composite with more functional groups and surface negative potential. *Sustain. Chem. Pharm.* **2021**, *24*, 100519. [CrossRef]
186. Subramani, S.E.; Thinakaran, N. Isotherm, kinetic and thermodynamic studies on the adsorption behaviour of textile dyes onto chitosan. *Process Saf. Environ. Prot.* **2017**, *106*, 1–10.
187. Zhang, W.; Huang, T.; Ren, Y.; Wang, Y.; Yu, R.; Wang, J.; Tu, Q. Preparation of chitosan crosslinked with metal-organic framework (MOF-199)@aminated graphene oxide aerogel for the adsorption of formaldehyde gas and methyl orange. *Int. J. Biol. Macromol.* **2021**, *193*, 2243–2251. [CrossRef]
188. Malek, N.N.A.; Jawad, A.H.; Ismail, K.; Razuan, R.; Alothman, Z.A. Fly ash modified magnetic chitosan-polyvinyl alcohol blend for reactive orange 16 dye removal: Adsorption parametric optimization. *Int. J. Biol. Macromol.* **2021**, *189*, 464–476. [CrossRef]
189. Shi, Y.; Song, G.; Li, A.; Wang, J.; Wang, H.; Sun, Y.; Ding, G. Graphene oxide-chitosan composite aerogel for adsorption of methyl orange and methylene blue: Effect of pH in single and binary systems. *Colloids Surf. A* **2022**, *641*, 128595. [CrossRef]

190. Liu, Y.; Li, L.; Duan, Z.; You, Q.; Liao, G.; Wang, D. Chitosan modified nitrogen-doped porous carbon composite as a highly-efficient adsorbent for phenolic pollutants removal. *Colloids Surf. A* **2021**, *610*, 125728. [CrossRef]
191. Zhang, L.; Sellaoui, L.; Franco, D.; Dotto, G.L.; Bajahzar, A.; Belmabrouk, H.; Bonilla-Petriciolet, A.; Oliveira, M.L.S.; Li, Z. Adsorption of dyes brilliant blue, sunset yellow and tartrazine from aqueous solution on chitosan: Analytical interpretation via multilayer statistical physics model. *Chem. Eng. J.* **2020**, *382*, 122952. [CrossRef]
192. Schio, R.R.; Gonçalves, J.O.; Mallmann, E.S.; Pinto, D.; Dotto, G.L. Development of a biosponge based on *Luffa cylindrica* and crosslinked chitosan for Allura red AC adsorption. *Int. J. Biol. Macromol.* **2021**, *192*, 1117–1122. [CrossRef] [PubMed]
193. Saheed, I.O.; Oh, W.-D.; Suah, F.B.M. Enhanced adsorption of acid Blue-25 dye onto chitosan/porous carbon composite modified in 1-allyl-3-methyl imidazolium bromide ionic liquid. *Int. J. Biol. Macromol.* **2021**, *183*, 1026–1033. [CrossRef] [PubMed]
194. Tang, T.; Cao, S.; Xi, C.; Chen, Z. Multifunctional magnetic chitosan-graphene oxide-ionic liquid ternary nanohybrid: An efficient adsorbent of alkaloids. *Carbohydr. Polym.* **2021**, *255*, 117338. [CrossRef]
195. Sahnoun, S.; Boutahala, M. Adsorption removal of tartrazine by chitosan/polyaniline composite: Kinetics and equilibrium studies. *Int. J. Biol. Macromol.* **2018**, *114*, 1345–1353. [CrossRef]
196. Daikh, S.; Ouis, D.; Benyoucef, A.; Mouffok, B. Equilibrium, kinetic and thermodynamic studies for evaluation of adsorption capacity of a new potential hybrid adsorbent based on polyaniline and chitosan for Acetaminophen. *Chem. Phys. Lett.* **2022**, 139565, in press. [CrossRef]
197. Pinheiro, C.P.; Moreira, L.M.K.; Alves, S.S.; Cadaval, T.R.S., Jr.; Pinto, L.A.A. Anthocyanins concentration by adsorption onto chitosan and alginate beads: Isotherms, kinetics and thermodynamics parameters. *Int. J. Biol. Macromol.* **2021**, *166*, 934–939. [CrossRef]
198. Turan, B.; Sarigol, G.; Demircivi, P. Adsorption of tetracycline antibiotics using metal and clay embedded cross-linked chitosan. *Mater. Chem. Phys.* **2022**, *279*, 125781. [CrossRef]
199. Jiang, Q.; Han, Z.; Li, W.; Ji, T.; Yuan, Y.; Zhang, J.; Zhao, C.; Cheng, Z.; Wang, S. Adsorption properties of heavy metals and antibiotics by chitosan from larvae and adult *Trypoxylus dichotomus*. *Carbohydr. Polym.* **2022**, *276*, 118735. [CrossRef]
200. Almeida, A.d.S.V.d.; Mastelaro, V.R.; da Silva, M.G.C.; Prediger, P.; Vieira, M.G.A. Adsorption of 17α-ethinylestradiol onto a novel nanocomposite based on graphene oxide, magnetic chitosan and organoclay (GO/mCS/OC): Kinetics, equilibrium, thermodynamics and selectivity studies. *J. Water Process Eng.* **2022**, *47*, 102729. [CrossRef]
201. Ranjbari, S.; Ayati, A.; Tanhaei, B.; Al-Othman, A.; Karimi, F. The surfactant-ionic liquid bi-functionalization of chitosan beads for their adsorption performance improvement toward Tartrazine. *Environ. Res.* **2022**, *204*, 111961. [CrossRef] [PubMed]
202. Yang, D.; Li, L.; Chen, B.; Shi, S.; Nie, J.; Ma, G. Functionalized chitosan electrospun nanofiber membranes for heavy-metal removal. *Polymer* **2019**, *163*, 74–85. [CrossRef]
203. Li, M.; Zhao, H.; Lu, Z.-Y. Porphyrin-based porous organic polymer, Py-POP, as a multifunctional platform for efficient selective adsorption and photocatalytic degradation of cationic dyes. *Microporous Mesoporous Mater.* **2020**, *292*, 109774. [CrossRef]
204. Wang, B.; Bai, Z.; Jiang, H.; Prinsen, P.; Luque, R.; Zhao, S.; Xuan, J. Selective heavy metal removal and water purification by microfluidically-generated chitosan microspheres: Characteristics, modeling and application. *J. Hazard. Mater.* **2019**, *364*, 192–205. [CrossRef] [PubMed]
205. Tang, S.; Yang, J.; Lin, L.; Peng, K.; Chen, Y.; Jin, S.; Yao, W. Construction of physically crosslinked chitosan/sodium alginate/calcium ion double-network hydrogel and its application to heavy metal ions removal. *Chem. Eng. J.* **2020**, *393*, 124728. [CrossRef]
206. Zhang, Z.; He, S.; Zhang, Y.; Zhang, K.; Wang, J.; Jing, R.; Yang, X.; Hu, Z.; Lin, X.; Li, Y. Spectroscopic investigation of Cu^{2+}, Pb^{2+} and Cd^{2+} adsorption behaviors by chitosan-coated argillaceous limestone: Competition and mechanisms. *Environ. Pollut.* **2019**, *254*, 112938. [CrossRef]
207. Hu, T.; Zeng, L.; Li, Y.; Wu, Y.; Zhu, Z.; Zhang, Y.; Tian, D.; Gao, C.; Li, W. Multifunctional chitosan non-woven fabrics modified with terylene carbon dots for selective detection and efficient adsorption of Cr(VI). *Chem. Eng. J.* **2022**, *432*, 134202. [CrossRef]
208. Zhu, X.; Tong, J.; Zhu, L.; Pan, D. In situ growth of ZIF-8 on carboxymethyl chitosan beads for improved adsorption of lead ion from aqueous solutions. *Int. J. Biol. Macromol.* **2022**, *205*, 473–482. [CrossRef]
209. Li, B.; Li, M.; Zhang, P.; Pan, Y.; Huang, Z.; Xiao, H. Remediation of Cd (II) ions in aqueous and soil phases using novel porous cellulose/chitosan composite spheres loaded with zero-valent iron nanoparticles. *React. Funct. Polym.* **2022**, *173*, 105210. [CrossRef]
210. Kameda, T.; Honda, R.; Kumagai, S.; Saito, Y.; Yoshioka, T. Adsorption of Cu^{2+} and Ni^{2+} by tripolyphosphate-crosslinked chitosan-modified montmorillonite. *J. Solid State Chem.* **2019**, *277*, 143–148. [CrossRef]
211. Lyu, F.; Yu, H.; Hou, T.; Yan, L.; Zhang, X.; Du, B. Efficient and fast removal of Pb^{2+} and Cd^{2+} from an aqueous solution using a chitosan/Mg-Al-layered double hydroxide nanocomposite. *J. Colloid Interface Sci.* **2019**, *539*, 184–193. [CrossRef] [PubMed]
212. Malwal, D.; Gopinath, P. Silica Stabilized Magnetic-Chitosan Beads for Removal of Arsenic from Water. *Colloid Interface Sci. Commun.* **2017**, *19*, 14–19. [CrossRef]
213. Marques Neto, J.d.O.; Bellato, C.R.; Silva, D. Iron oxide/carbon nanotubes/chitosan magnetic composite film for chromium species removal. *Chemosphere* **2019**, *218*, 391–401. [CrossRef]
214. Neeraj, G.; Krishnan, S.; Senthil Kumar, P.; Shriaishvarya, K.R.; Vinoth Kumar, V. Performance study on sequestration of copper ions from contaminated water using newly synthesized high effective chitosan coated magnetic nanoparticles. *J. Mol. Liq.* **2016**, *214*, 335–346. [CrossRef]

215. Periyasamy, S.; Manivasakan, P.; Jeyaprabha, C.; Meenakshi, S.; Viswanathan, N. Fabrication of nano-graphene oxide assisted hydrotalcite/chitosan biocomposite: An efficient adsorbent for chromium removal from water. *Int. J. Biol. Macromol.* **2019**, *132*, 1068–1078. [CrossRef] [PubMed]
216. Nematidil, N.; Sadeghi, M.; Nezami, S.; Sadeghi, H. Synthesis and characterization of Schiff-base based chitosan-g-glutaraldehyde/NaMMTNPs-APTES for removal Pb^{2+} and Hg^{2+} ions. *Carbohydr. Polym.* **2019**, *222*, 114971. [CrossRef] [PubMed]
217. Xiong, Y.; Xie, L.; Zhu, L.; Wang, Y.; Shan, W.; Lou, Z.; Cui, J.; Yu, H. Superior adsorption of Re(VII) by anionic imprinted chitosan-silica composite: Adsorption performance, selectivity and mechanism study. *J. Ind. Eng. Chem.* **2022**, *108*, 344–355. [CrossRef]
218. Yang, L.; Luo, X.; Yan, L.; Zhou, Y.; Yu, S.; Ju, H.; Wang, Y.; Zhang, L. Efficient selective adsorption of uranium using a novel eco-friendly chitosan-grafted adenosine 5′-monophosphate foam. *Carbohydr. Polym.* **2022**, *285*, 119157. [CrossRef]
219. Ding, W.; Zhang, J.; Liu, Y.; Guo, Y.; Deng, T.; Yu, X. Synthesis of granulated H4Mn5O12/chitosan with improved stability by a novel cross-linking strategy for lithium adsorption from aqueous solutions. *Chem. Eng. J.* **2021**, *426*, 131689. [CrossRef]
220. Liu, M.; Zang, Z.; Zhang, S.; Ouyang, G.; Han, R. Enhanced fluoride adsorption from aqueous solution by zirconium (IV)-impregnated magnetic chitosan graphene oxide. *Int. J. Biol. Macromol.* **2021**, *182*, 1759–1768. [CrossRef]
221. Yang, L.; Huang, C.; Luo, X.; Zhang, L.; Ye, Y.; Jun, H.; Wang, Y. Chitosan-based aerogel with anti-swelling for U(VI) adsorption from aqueous solution. *Colloids Surf. A* **2021**, *630*, 127527. [CrossRef]
222. Wang, S.; Wang, H.; Tang, J.; Chen, Y.; Wang, S.; Zhang, L. Chitosan functionalized with N,N-(2-aminoethyl)pyridinedicarboxamide for selective adsorption of gold ions from wastewater. *Int. J. Biol. Macromol.* **2022**, *194*, 781–789. [CrossRef] [PubMed]
223. Laureano-Anzaldo, C.M.; González-López, M.E.; Pérez-Fonseca, A.A.; Cruz-Barba, L.E.; Robledo-Ortíz, J.R. Synthesis of silanized chitosan anchored onto porous composite and its performance in fixed-bed adsorption of Cr(VI). *J. Environ. Chem. Eng.* **2021**, *9*, 106353. [CrossRef]
224. Gamal, A.; Ibrahim, A.G.; Eliwa, E.M.; El-Zomrawy, A.H.; El-Bahy, S.M. Synthesis and characterization of a novel benzothiazole functionalized chitosan and its use for effective adsorption of Cu(II). *Int. J. Biol. Macromol.* **2021**, *183*, 1283–1292. [CrossRef] [PubMed]
225. Sharma, G.; Naushad, M.; Ala'a, H.; Kumar, A.; Khan, M.R.; Kalia, S.; Bala, M.; Sharma, A. Fabrication and characterization of chitosan-crosslinked-poly(alginic acid) nanohydrogel for adsorptive removal of Cr(VI) metal ion from aqueous medium. *Int. J. Biol. Macromol.* **2017**, *95*, 484–493. [CrossRef] [PubMed]
226. Chen, Y.; Tang, J.; Wang, S.; Zhang, L. Ninhydrin-functionalized chitosan for selective removal of Pb(II) ions: Characterization and adsorption performance. *Int. J. Biol. Macromol.* **2021**, *177*, 29–39. [CrossRef]
227. Zhuang, S.; Zhu, K.; Xu, L.; Hu, J.; Wang, J. Adsorption of Co^{2+} and Sr^{2+} in aqueous solution by a novel fibrous chitosan biosorbent. *Sci. Total Environ.* **2022**, *825*, 153998. [CrossRef]
228. Chen, L.; Tang, J.; Zhang, X.; Wang, S.; Ren, Z. A novel benzothiazole modified chitosan with excellent adsorption capacity for Au(III) in aqueous solutions. *Int. J. Biol. Macromol.* **2021**, *193*, 1918–1926. [CrossRef]
229. Huang, H.; Yang, Q.; Zhang, L.; Huang, C.; Liang, Y. Polyacrylamide modified kaolin enhances adsorption of sodium alginate/carboxymethyl chitosan hydrogel beads for copper ions. *Chem. Eng. Res. Des.* **2022**, *180*, 296–305. [CrossRef]
230. Huang, Y.; Wu, Y.; Ding, W.; Sun, Q.; Hu, C.; Liu, B.; Liu, H.; Zheng, H. Anion-synergistic adsorption enhances the selective removal of silver ions from complex wastewater by chitosan-coated magnetic silica core-shell nanoparticles. *J. Cleaner Prod.* **2022**, *339*, 130777. [CrossRef]
231. Ding, Y.; Liu, D.; Luo, D.; Sun, X.; Mei, J.; Wang, S.; Li, Z. Rapid one-step preparation of a carboxymethyl chitosan gel with a novel crosslinker for efficient adsorption of Sr^{2+}. *Colloids Surf. A* **2022**, *641*, 128576. [CrossRef]
232. Zeng, H.; Xu, K.; Wang, F.; Sun, S.; Li, D.; Zhang, J. Adsorption of As(III) from aqueous solutions using MnO_2 strengthened WTRs-chitosan beads made by homogenous method with freeze-drying. *React. Funct. Polym.* **2021**, *167*, 105016. [CrossRef]
233. Li, H.; Ji, H.; Cui, X.; Che, X.; Zhang, Q.; Zhong, J.; Jin, R.; Wang, L.; Luo, Y. Kinetics, thermodynamics, and equilibrium of As(III), Cd(II), Cu(II) and Pb(II) adsorption using porous chitosan bead-supported $MnFe_2O_4$ nanoparticles. *Int. J. Min. Sci. Technol.* **2021**, *31*, 1107–1115. [CrossRef]
234. Abugoch, L.E.; Tapia, C.; Villamán, M.C.; Yazdani-Pedram, M.; Díaz-Dosque, M. Characterization of quinoa protein–chitosan blend edible films. *Food Hydrocoll.* **2011**, *25*, 879–886. [CrossRef]
235. Salam, A.; Lucia, L.A.; Jameel, H. Synthesis, characterization, and evaluation of chitosan-complexed starch nanoparticles on the physical properties of recycled paper furnish. *ACS Appl. Mater. Interfaces* **2013**, *5*, 11029–11037. [CrossRef]
236. Balea, A.; Monte, M.C.; Fuente, E.; Sanchez-Salvador, J.L.; Blanco, A.; Negro, C. Cellulose nanofibers and chitosan to remove flexographic inks from wastewaters. *Environ. Sci. Water Res. Technol.* **2019**, *5*, 1558–1567. [CrossRef]
237. Rahmaninia, M.; Rohi, M.; Hubbe, M.A.; Zabihzadeh, S.M.; Ramezani, O. The performance of chitosan with bentonite microparticles as wet-end additive system for paper reinforcement. *Carbohydr. Polym.* **2018**, *179*, 328–332. [CrossRef]
238. Todorova, D.; Lasheva, V. Effect of Chitosan Addition during Paper-Making on Ageing Stability of Document Paper. *Cellul. Chem. Technol.* **2021**, *55*, 1083–1094. [CrossRef]
239. Ashori, A.; Harun, J.; Zin, W.M.; Yusoff, M.N.M. Enhancing Dry-Strength Properties of Kenaf (*Hibiscus cannabinus*) Paper Through Chitosan. *Polym. Plast. Technol. Eng.* **2006**, *45*, 125–129. [CrossRef]

240. Kasim, N.F.A.; WIdris, W.F.; Abdullah, A.H.; Yusoh, K.; Ismail, Z. The preparation of graphene ink from the exfoliation of graphite in pullulan, chitosan and alginate for strain-sensitive paper. *Int. J. Biol. Macromol.* **2020**, *153*, 1211–1219. [CrossRef]
241. Khoushab, F.; Yamabhai, M. Chitin research revisited. *Mar. Drugs* **2010**, *8*, 1988–2012. [CrossRef] [PubMed]
242. Azharul Islam, M.; Tan, Y.L.; Atikul Islam, M.; Romić, M.; Hameed, B.H. Chitosan–bleaching earth clay composite as an efficient adsorbent for carbon dioxide adsorption: Process optimization. *Colloids Surf. A* **2018**, *554*, 9–15. [CrossRef]
243. Hsan, N.; Dutta, P.K.; Kumar, S.; Koh, J. Arginine containing chitosan-graphene oxide aerogels for highly efficient carbon capture and fixation. *J. CO_2 Util.* **2022**, *59*, 101958. [CrossRef]
244. Mincke, S.; Asere, T.G.; Verheye, I.; Folens, K.; Vanden Bussche, F.; Lapeire, L.; Verbeken, K.; Van Der Voort, P.; Tessema, D.A.; Fufa, F.; et al. Functionalized chitosan adsorbents allow recovery of palladium and platinum from acidic aqueous solutions. *Green Chem.* **2019**, *21*, 2295–2306. [CrossRef]
245. Borgohain, R.; Jain, N.; Prasad, B.; Mandal, B.; Su, B. Carboxymethyl chitosan/carbon nanotubes mixed matrix membranes for CO_2 separation. *React. Funct. Polym.* **2019**, *143*, 104331. [CrossRef]
246. Kamran, U.; Park, S.-J. Tuning ratios of KOH and NaOH on acetic acid-mediated chitosan-based porous carbons for improving their textural features and CO_2 uptakes. *J. CO_2 Util.* **2020**, *40*, 101212. [CrossRef]
247. Rehman, A.; Park, S.-J. From chitosan to urea-modified carbons: Tailoring the ultra-microporosity for enhanced CO_2 adsorption. *Carbon* **2020**, *159*, 625–637. [CrossRef]
248. Hsan, N.; Dutta, P.K.; Kumar, S.; Das, N.; Koh, J. Capture and chemical fixation of carbon dioxide by chitosan grafted multi-walled carbon nanotubes. *J. CO_2 Util.* **2020**, *41*, 101237. [CrossRef]

polymers

Article

Synthesis and Characterization of Biodegradable Poly(butyl cyanoacrylate) for Drug Delivery Applications

Benjamin-Luca Keller [1,*], Claudia A. Lohmann [2,*], Samuel O. Kyeremateng [1] and Gert Fricker [3]

1 AbbVie Deutschland GmbH & Co. KG, Knollstrasse, 67061 Ludwigshafen, Germany; samuel.kyeremateng@abbvie.com
2 Waters Corporation, 34 Maple Street, Milford, MA 01757, USA
3 Institute of Pharmacy and Molecular Biotechnology, University of Heidelberg, Im Neuenheimer Feld 329, 69120 Heidelberg, Germany; gert.fricker@uni-hd.de
* Correspondence: benjamin-luca.keller@abbvie.com (B.-L.K.); claudia_lohmann@waters.com (C.A.L.); Tel.: +49-621-589-2392 (B.-L.K.); +1-508-482-2305 (C.A.L.)

Abstract: Poly(butyl cyanoacrylate) (PBCA) is a biodegradable and biocompatible homopolymer which is used as a carrier matrix for drug delivery systems in the pharmaceutical industry. Typically, polymerization is carried out under aqueous conditions and results in molecular weights are mostly lower than 3000 g/mol due to the instability of the high molecular weight PBCA. However, the stability of polymer excipients is a major prerequisite for drug product development in the pharmaceutical industry. In this work, a reliable polymer synthesis strategy for PBCA was designed to control the molecular weight in a nonaqueous polymerization environment. The anionic polymerization process and the impact of key synthesis parameters were investigated. The results confirmed that the previously postulated depolymerization–repolymerization process (DPRP) in the literature can be used to tailor the molecular weight of PBCA. The amount of sodium methoxide present during the polymerization proved to be the key parameter to control the DPRP and the molecular weight as desired. In addition, it was discovered that end-capping the PBCA chain suppressed the DPRP and prevented monomer release by depriving the PBCA of its living character. Thus, neat PBCA polymer with varying molecular weights determined by Advanced Polymer Chromatography™ as well as end-capped PBCA were synthesized, and the improvement of the chemical and shelf-life stability were confirmed using NMR.

Keywords: drug delivery; PBCA; molecular weights; biodegradable polymers; NMR; Advanced Polymer Chromatography™

Citation: Keller, B.-L.; Lohmann, C.A.; Kyeremateng, S.O.; Fricker, G. Synthesis and Characterization of Biodegradable Poly(butyl cyanoacrylate) for Drug Delivery Applications. *Polymers* **2022**, *14*, 998. https://doi.org/10.3390/polym14050998

Academic Editors: Florian J. Stadler and Alberto García-Peñas

Received: 20 January 2022
Accepted: 26 February 2022
Published: 28 February 2022

Publisher's Note: MDPI stays neutral with regard to jurisdictional claims in published maps and institutional affiliations.

1. Introduction

Over the past few decades, nanomedicine has become a promising drug delivery strategy which provides solutions to increase drug substance concentrations at the site of action. The term nanomedicine is associated with the approach to target specific sites of action within a patient with drug-loaded nanoparticles (NP). These include polymeric NP, metal or metal oxide NP, nanocrystals, carbon-based NP, protein NP, lipid NP and liposomes [1–8]. Polymer-based NP have been known and investigated for the past four decades. In particular, polymeric NP comprised of biodegradable polymers derived from synthetic or natural polymers offer a promising approach for targeted drug delivery. Commonly used biodegradable and biocompatible polymers for polymeric NP design include poly(alkyl cyanoacrylates) (PACA) [9]. In 1979, Couvreur et al. reported the generation and characterization of PACA-based NP for the first time [10]. Since then, PACA NP have been extensively studied and frequently published [11,12]. Different PACA types with varying side alkyl groups up to C8 find use for NP production [12]. In the pharmaceutical field, poly(butyl cyanoacrylate) (PBCA) is a common representative of the PACA family, frequently studied as a drug delivery system to target cancer and/or pass the blood–brain barrier [11,13]. PBCA is a homopolymer composed of *n*-butyl-2-cyanoacrylate (BCA) monomer units. The monomer is successfully applied as a biomedical tissue adhesive due to the fast polymerization mechanism [14]. However, the lack of commercially available PBCA bears a challenge for the production of PBCA-based drug delivery systems. Thus, PBCA must be synthesized from the monomer prior to NP production. This distinguishes PBCA NP considerably from other polymeric NP systems (e.g., poly(lactic-co-glycolide) or poly(caprolactone)), where the polymer matrices are commercially available.

PBCA can generally be considered a pharmaceutical excipient; therefore, it has to meet the requirements regarding safety, functionality and quality [15,16]. The lack of commercially available PBCA has so far been compensated by applying BCA polymerization-based NP production methods where the in situ formation of polymer and nanoparticles take place simultaneously [11,12]. Within this approach, the anionic living polymerization is the predominant polymerization method for the synthesis of PBCA. The anionic polymerization of PBCA and its mechanism was intensively studied by Pepper et al. [17]. Although this polymerization-based method is mainly used to generate PBCA NP in the pharmaceutical field, the one-step process has two main drawbacks:

1. The characterization of PBCA as a defined pharmaceutical excipient to ensure quality prior to the NP formation is not possible;
2. The solvent-dependent molecular weight (MW) control insufficiency of PCBA during polymerization.

Within the pharmaceutical field, BCA is usually polymerized under aqueous conditions. In an aqueous medium, the formation of oligomeric units is favored, yielding a MW range of 1000–3000 g/mol [18]. Based on the literature, no polymeric NP-based delivery systems composed of PBCA exceed a MW of 3000 g/mol. However, in the field of polymeric NP, PBCA with higher MW is of great interest, since it was shown that the MW of biodegradable polymers can influence the physicochemical properties of the NP and thereby impact their performance after administration [19–21].

The reason for the insufficient polymerization control in an aqueous environment can be attributed to the high reactivity of the monomer's chemical structure. The β-carbonyl-like structure of alkyl cyanoacrylates has a preferred tendency to form carbanions which undergo resonance stabilization induced by the strong electronegativity of the nitrile and carbonyl group (Figure 1A). This potent electron-withdrawing tendency increases the terminal C-H acidity as the chain length of the polymer increases (Figure 1B). Carbanion stabilization lends an exceptionally high reactivity to the monomer. The C-H acidity contributes to the living character of a PBCA molecule, enabling the polymer chain to restructure depending on the surrounding environment [18].

However, Pepper et al. showed that it is also possible to perform a living polymerization of PBCA in anhydrous THF as a polymerization medium [17,22]. Trace amounts of anions and weak bases such as hydroxides, alcohols, amines, etc., are sufficient to initiate the polymerization of PBCA. Furthermore, the researchers discovered a polymerization-inhibiting effect of strong acids that could also retard the polymerization of BCA [23,24]. The inhibition is caused by an excess of protons neutralizing the number of basic molecules in the reaction mixture as well as carbanion formation by covalently bonding to the α-carbon. Other possible polymerization mechanisms for PBCA are zwitterionic and radical polymerization. These are rarely performed in the pharmaceutical industry [25–27] because anionic polymerization always occurs as a side reaction.

Putting a strong focus on developing a reliable synthesis process for PBCA as an excipient calls for separating the synthesis process from the nanoparticle formation process in order to independently control the MW of PBCA. Therefore, in this work, the anionic polymerization mechanism of PBCA was studied and key polymerization process parameters were investigated and varied. The evolution of the polymerization process was monitored by Advanced Polymer Chromatography™ (APC™), determining the MW of the polymer at different stages of the polymerization process. APC™ is an UPLC-like instrument that is utilized for size-based separations, offering the unique capability of high-resolution analysis. Due to its ability to resolve low MW species down to the monomer level, APC™ proves to be a valuable tool for analyzing the changes during polymerization. The short runtimes of a single measurement also allow for a high sample throughput. NMR was used to verify the chemical structure of PBCA and monitor the postsynthesis end-capping functionalization of PBCA.

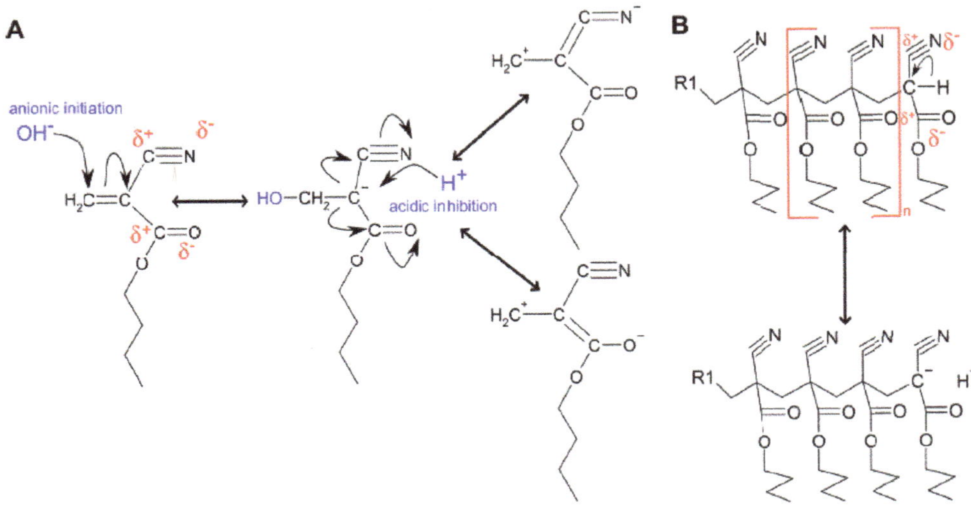

Figure 1. (**A**) Chemical structure of BCA, including possible resonance structures; (**B**) Chemical structures of PBCA and the origin of the C-H acidity appearing on the chain terminus of the polymer.

The overall aim of this work was to control and optimize the anionic polymerization process for PBCA to allow tailoring of the polymer MW. Keeping in mind that PBCA can potentially be applied as a pharmaceutical excipient for drug delivery, we further investigated the MW effect on the chemical stability of PBCA and end-capped PBCA.

2. Materials and Methods

2.1. Chemicals

Indermil (*n*-butyl cyanoacrylate monomer, BCA) was obtained from Henkel (Düsseldorf, Germany). AcroSeal™ tetrahydrofuran (THF) 99.5% extra dry was purchased from Acros Organics (Geel, Belgium) and trifluoroacetic acid (TFA) was purchased from Omnilab (Bremen, Germany). Sodium methoxide (NaOMe) powder, deuterated chloroform ($CDCl_3$) and ethyl 2-(bromomethyl) acrylate (EBMA) were obtained from Sigma-Aldrich (Taufkirchen, Germany). A total of 0.1 M hydrochloric acid (HCl), chloroform (HPLC grade), methanol (HPLC grade), acetonitrile (HPLC grade) and HPLC-grade water were purchased from Merck (Darmstadt, Germany).

2.2. Characterization Techniques

2.2.1. Advanced Polymer Chromatography™ (APC™)

ACQUITY™ APC™ from Waters Corporation (Milford, MA, USA) was used with a column bank consisting of an ACQUITY™ BEH XT 200 (4.6 mM × 150 mM, 2.5 µm) connected to two ACQUITY™ BEH XT 45 (4.6 mM × 150 mM, 1.7 µm) in series at 40 °C. THF was used as a mobile phase at a flow rate of 1 mL/min. Refractive index was used as a detector for the chromatograms. Conventional calibration was conducted with polystyrene standards. MW calculations were conducted with Empower™ with GPC option. Samples were dissolved in THF to make a final concentration of 5 mg/mL and 10 µL of the solution injected for the measurements.

2.2.2. NMR

The 600 MHz ^{1}H and 150 MHz ^{13}C NMR spectra were obtained with a Bruker Avance™ system (Karlsruhe, Germany). Approximately 5 mg of samples were dissolved in 600 µL deuterated chloroform ($CDCl_3$) for the measurements.

2.3. Synthesis

2.3.1. Anionic Polymerization in THF

TFA diluted in 5 mL of THF was mixed with BCA in 20 mL scintillation vials. A total 120 µL of an aqueous solution of NaOMe was added under continuous magnetic stirring according to Table 1. Stirring at room temperature was continued for 180 min. Within this period, samples were taken at different time points and the polymerization was quenched immediately by precipitation in 0.1 M HCl (pH 1). The precipitates were frozen on dry ice and then lyophilized in an Alpha LSC 2–4 freeze dryer from Martin Christ (Osterode am Harz, Germany). Primary drying in the freeze dryer was performed for 12 h at 0.37 mbar with a shelf temperature of 0 °C. Secondary drying was done for 1 h at 0.01 mbar with a shelf temperature of 10 °C.

Table 1. Conditions during anionic polymerization in THF.

Condition	TFA (mM)	BCA (mM)	NaOMe (mM)
1	0.00	326	0
2	0.01	326	5
3	1.00	326	5
4	1.00	326	8
5	1.00	326	100
6	1.00	653	8
7	2.00	326	0.125
8	2.00	326	0.630
9	2.00	326	10
10	2.00	326	20
11	2.00	326	40
12	2.00	326	75
13	2.00	326	80

2.3.2. PBCA End-Capping with EBMA in THF

A total of 2 mM TFA diluted in 5 mL THF was mixed with a 326 mM BCA solution under continuous magnetic stirring. The polymerization was started by adding 120 µL of an aqueous 8 mM or 80 mM NaOMe solution. Stirring was continued for 3 h. An amount of 2.5 mL of a 5 mM or 25 mM EBMA solution in THF was mixed with 2.5 mL of the PBCA polymerization mixture. The EBMA–PBCA mixture was stirred for 1 h. The EBMA-terminated PBCA (tPBCA) was precipitated in 0.1 M HCl (pH 1), purified by filtration and washed with deionized water to remove excess EBMA and lyophilized as described in Section 2.3.1.

2.3.3. Polymer Stability in Solution

Approximately 5 mg polymer was dissolved in 600 µL deuterated chloroform and stored in a closed glass vial for seven weeks at room temperature. A 600 MHz ^{1}H NMR spectrum was recorded each week.

3. Results

3.1. Polymerization Process Monitoring by APC™

Following Ryan et al.'s, approach, anionic polymerization was performed in THF [22]. The polymerization design comprised of reacting varying amounts of NaOMe, BCA and

TFA. The evaluated polymerization conditions are reported in Table 1. The APC™ was used to monitor the evolution of the polymerization process and to investigate the impact of the polymerization condition on the MW of the target PBCA.

As described in Section 2, the BCA was diluted in THF containing TFA, which was used as a polymerization-retarding component. In general, the literature reports acids as polymerization-inhibiting components for the anionic polymerization of BCA [19,24]. Protons delivered by acids occupy the C-H acidic unit on the BCA structure. Consequently, carbanion formation is prevented and, therewith, the polymerization initiation. Costa et al. distinguished between the "polymerization inhibition" induced by strong acids (such as H_2SO_4 or CF_3SO_3H) and the "polymerization retardations periods" induced by weaker acids (such as $ClCH_2COOH$) [24]. Thus, the presence of TFA prevented the spontaneous polymerization in the solvent THF before an initiator was added to the BCA–THF mixture. The presence of an acid is necessary because BCA has the tendency to self-initiate spontaneous polymerization, particularly in polar solvents such as THF [26].

The addition of different NaOMe amounts as the polymerization initiator led to the formation of mainly high molecular weight (HMW) PBCA, referred to as the "parent peak", within a minute, as illustrated in Figure 2. Alongside the main HMW population is the low molecular weight (LMW) population, referred to as "daughter peak", as well as a minor population at high retention times. Figure 2 also depicts the time-dependent emergence or growth of the "daughter peak" at the expense of the "parent peak" until all of the latter is consumed.

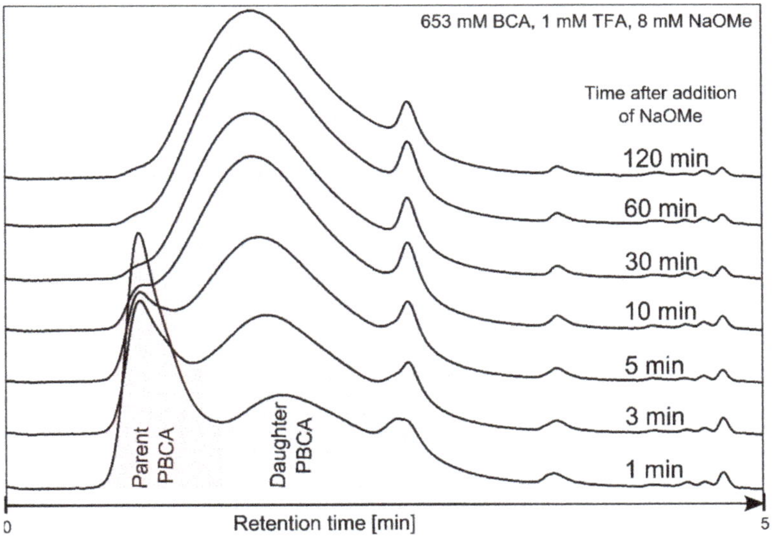

Figure 2. Time-dependent APC™ chromatograms monitoring the polymerization of PBCA following NaOMe addition.

This indicates that the initial HMW PBCA underwent depolymerization, exhibited as decrease in the peak intensity and area. Thus, the depolymerization of the parent polymer was accompanied by a simultaneous repolymerization of the unzipped monomer. The distinctive formation, as well as the complete degeneration of the parent PBCA, was strongly dependent on selected polymerization conditions. The restructuring of PBCA from parent to daughter polymer, or the depolymerization–repolymerization process (DPRP), and the subsequent MW changes and relative peak areas are listed in Table 2. Similar observations were first reported by Ryan et al. [22]. The mechanism of the postulated DPRP is displayed in Figure 3.

Table 2. MW and relative peak area of parent and daughter PBCA as determined by APC™ (653 mM BCA, 1 mM TFA and 8 mM NaOMe).

Time (min)	Parent PBCA		Daughter PBCA	
	MW (g/mol)	Relative Peak Area (%)	MW (g/mol)	Relative Peak Area (%)
0.5	181,488	67	645	33
1	127,605	49	12,023	51
3	131,730	27	15,751	73
5	145,970	15	18,775	85
10	168,314	4	21,623	96
30	-	0	23,328	100
60	-	0	22,496	100
120	-	0	22,388	100
180	-	0	21,713	100

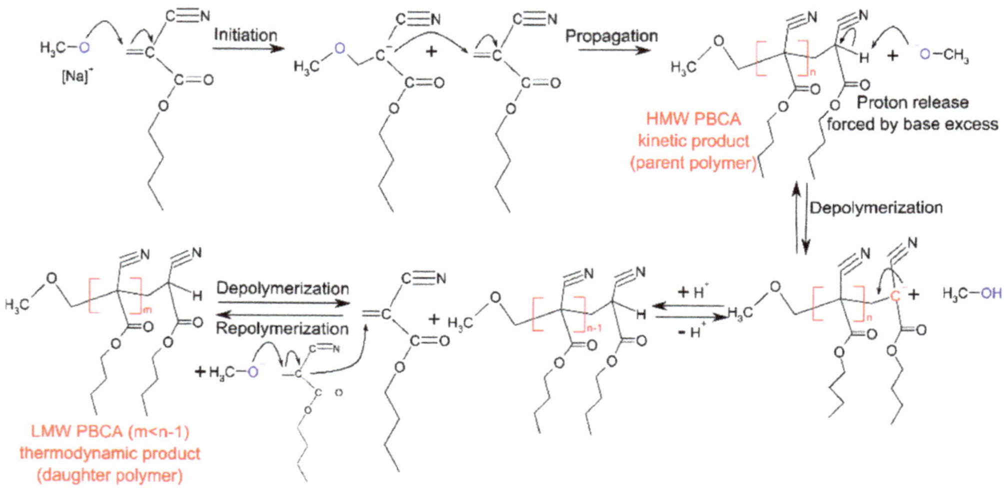

Figure 3. Initiation, propagation and depolymerization–repolymerization steps involved during anionic polymerization of BCA initiated by NaOMe in THF.

From a postulated mechanistic standpoint, the initial and rate determining step of the DPRP process is the proton liberation from the terminal unit in the polymer chain [22,28]. This proton release induces monomer unzipping from the chain terminus. Ryan et al. concluded from their studies that the excess base can facilitate the proton release and subsequently induce the formation of the LMW daughter PBCA [29]. The excess base forced the polymerization towards a thermodynamic product by deprotonation of the terminal polymer unit and thus reduced the activation energy of the rate-determining step. The researchers described the parent polymer as a kinetic product and the daughter polymer as a thermodynamic product. However, many of the investigated DPRP process conditions in this work resulted in a fluent transition from HMW to LMW PBCA without the appearance of two clearly separated parent and daughter fractions in the APC™ chromatogram, as can be seen in Figure 4. These transitions, along with the MW changes, appeared as a peak shift in the chromatogram from small retention times resembling HMW parent fractions to high retention times for the LMW daughter fractions.

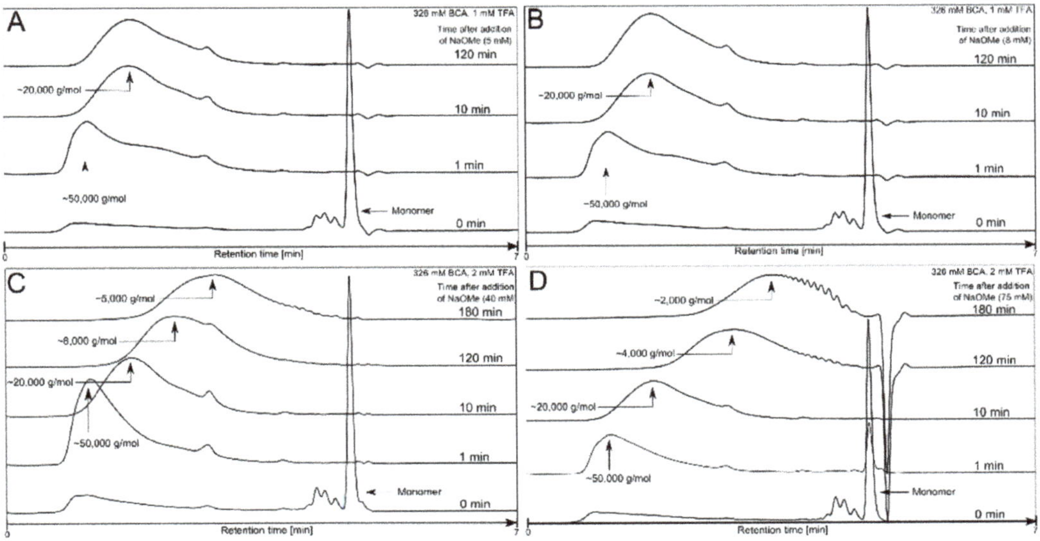

Figure 4. APC™ chromatogram of PBCA at different polymerization time points obtained with 5 mM NaOMe (**A**), 8 mM NaOMe (**B**), 40 mM NaOMe (**C**) and 75 mM NaOMe (**D**).

The results in Figures 5 and 6 indicate that the DPRP of PBCA is affected by the amounts of NaOMe, BCA and TFA. NaOMe at varying amounts had the most profound impact on the DPRP. With increasing amounts of NaOMe, the MW decreased significantly within 180 min of polymerization. The trend of decreasing MW dependent on the NaOMe amount was observed in all polymerization conditions, as detailed in Figure 5.

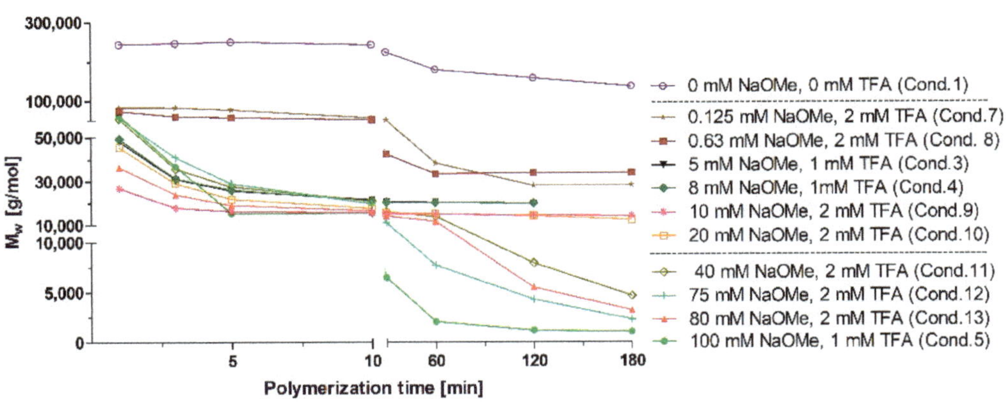

Figure 5. MW changes of PBCA during the anionic polymerization process induced by different NaOMe amounts (0–100 mM). A total of 326 mM BCA was used in all plotted conditions. MW was determined by APC™.

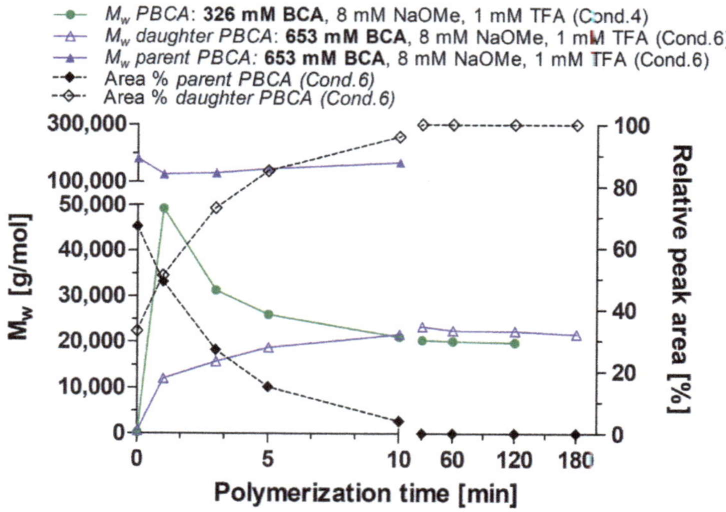

Figure 6. Influence of BCA amount on MW changes during the polymerization process. Comparison between 326 mM BCA and 653 mM BCA, both with 1 mM TFA and 8 mM NaOMe. MW was determined by APC™.

The outcome of the trend plots (Figures 5 and 6) clearly supports previous studies reporting that the DPRP can be influenced by basic molecules such as TBOH or NaOMe [22,30,31]. The results demonstrate that the effect on the PBCA MW depends on the NaOMe addition and supports the postulated reaction mechanism of the DPRP in Figure 3. Generally, it can be assumed that any base which can facilitate the proton release will equally reduce the activation energy and consequently force the DPRP towards LMW. Furthermore, different amounts of monomer with fixed amounts of TFA and NaOMe affected the polymerization process. An overall trend plot displaying the MW change induced by varying amounts of BCA is depicted in Figure 6. Polymerization with 326 mM BCA, 8 mM NaOMe and 1 mM TFA led to a fluent PBCA transition without the occurrence of a parent PBCA fraction.

In contrast, increasing the BCA amount (653 mM) decelerated the DPRP, which led to the distinct formation of parent PBCA and daughter PBCA fractions during the polymerization. It can be assumed that a larger monomer amount (653 mM) generated a larger number of HMW PBCA molecules. Hence, under similar conditions, more time is required to depolymerize the large number of HMW fractions compared to the polymerization with 326 mM BCA. Consequently, an increased amount of HMW PBCA molecules will extend the time for the total HMW PBCA consumption and therewith increase the transformation time of HMW PBCA to LMW PBCA. In this case, the HMW PBCA consumption delay prevented a fluent transition and led to the appearance of both fractions in the APC™ chromatogram. However, the final MW is similar (20,000 g/mol) in both the 326 mM and 653 mM BCA trials because the key factor for the final MW at the end of the polymerization is the NaOMe amount, which was kept constant at 8 mM (Figure 6).

3.2. NMR

3.2.1. PBCA Structure Elucidation

Two PBCA batches with different MW (2000 g/mol and 20,000 g/mol) were analyzed by ^{13}C and ^{1}H NMR to verify the structure (Figure 7). Additionally, ^{1}H and ^{13}C spectra of the BCA monomer were recorded to identify potential residual monomer units. Both NMR measurements successfully confirmed the PBCA structure [30,32–35]. Differences between the 2000 g/mol and 20,000 g/mol PBCA batches in the ^{13}C and ^{1}H spectra appeared to be negligible. Therefore, the NaOMe amounts (8 mM or 80 mM) added to control the MW did not affect the PBCA structure from a chemical perspective. Both spectra revealed some unexpected signals (marked as * in Figure 7) originating from butylated hydroxytoluene (BHT) that stems from the THF polymerization medium. The BHT-laced polymers are indicative of an insufficient purification process at this lab scale and may require optimization in case of scale up. Alternatively, unstabilized BHT-free THF can be utilized instead.

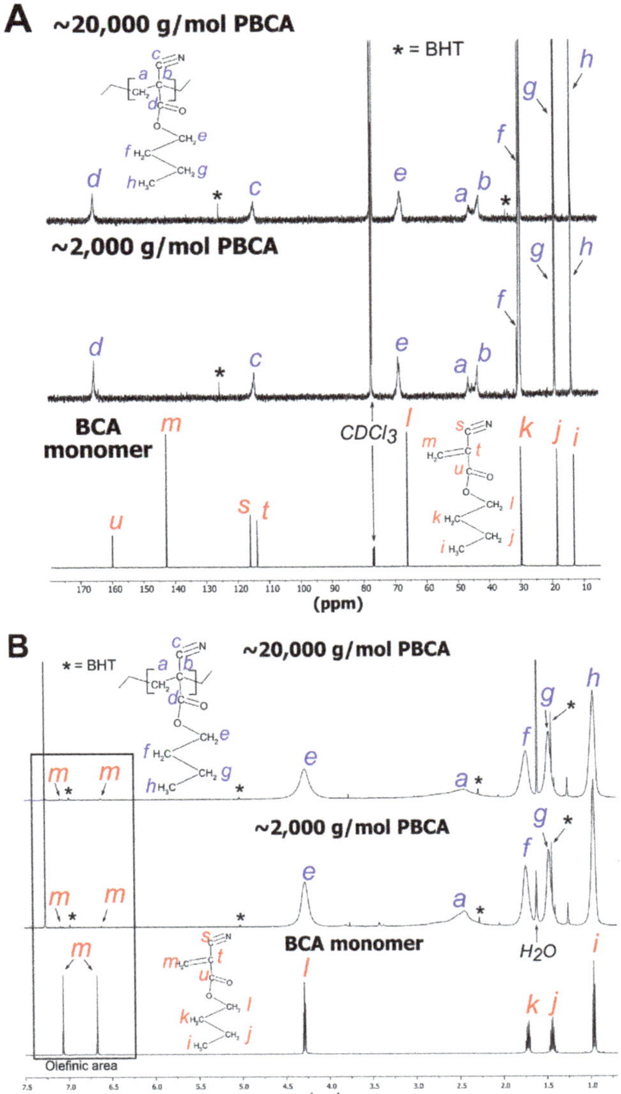

Figure 7. (**A**) The 150 MHz ^{13}C NMR spectra of 2000 g/mol PBCA, 20,000 g/mol PBCA and BCA monomer with molecular structure peak assignments; (**B**) The 600 MHz ^{1}H NMR spectra of 2000 g/mol PCBA, 20,000 g/mol PBCA and monomer with molecular structure peak assignments. Signals associated with butylated hydroxytoluene (BHT) impurity are marked as *.

Signal band broadening in the spectra suggest PBCA to be an atactic polymer (Figure 7A). In addition, the ^{1}H spectra of the 2000 g/mol and 20,000 g/mol PBCA batches showed very weak signals (m) in the olefinic area at 7.06 ppm and 6.62 ppm. Comparing the polymer spectra with the monomer spectrum confirmed that these olefinic signals originated from the monomeric vinyl group (m) (Figure 7B). Similar observations of minor amounts of residual monomer (m) in the ^{1}H spectra in the polymer were also made by other research groups after the polymerization of the alkyl cyanoacrylates [33,36]. The residual monomer amount was negligible and only detectable due to the low detection limit of the ^{1}H NMR measurement. In the ^{13}C NMR measurements, it was not possible to detect the residual monomer due to the lower sensitivity.

3.2.2. tPBCA Structure Elucidation

For the systematic investigation of the living polymer character of PBCA, and potentially improving the stability of the PBCA polymer, the C-H acidic end group was substituted by EBMA in a postpolymerization reaction, as described by Kohsaka et al. [37]. Briefly, the EBMA acts as an efficient terminator by attacking the living carbanion of PBCA on the vinylidene group of the EBMA, with the subsequent elimination of bromine to generate an unsaturated ester moiety at the polymer end chain. Consequently, the termination with EBMA deprives the PBCA chain of its living character.

Two different MW batches of PBCA with a MW of 2000 g/mol and 20,000 g/mol were terminated at two EBMA concentrations, i.e., 25 mM or 5 mM. The structure of tPBCA was analyzed by NMR. The ^{13}C and ^{1}H spectra of tPBCA confirmed the successful linkage of the polymer and the terminal EBMA molecule (Figure 8).

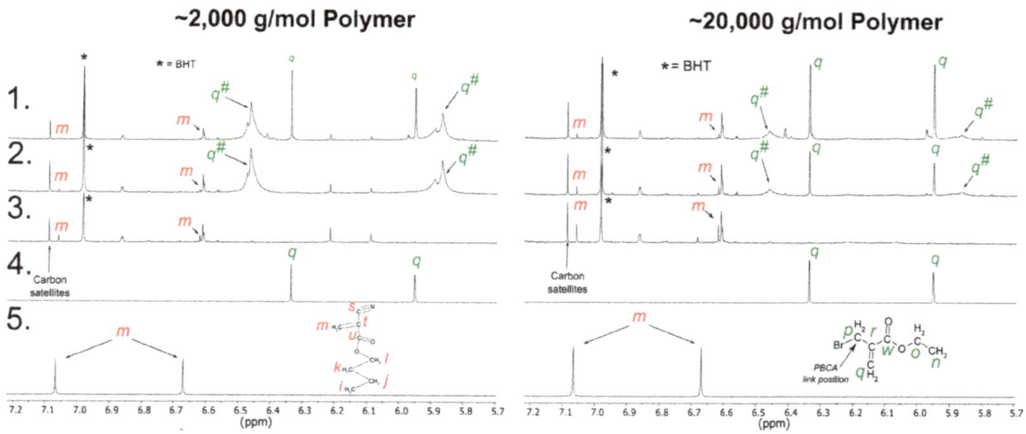

Figure 8. The 600 MHz ^{1}H NMR spectra of the olefinic region. (**Left**): ~2000 g/mol tPBCA using 25 mM EBMA as terminator (**1**), ~2000 g/mol tPBCA using 5 mM EBMA (**2**), ~2000 g/mol PBCA as comparison (**3**), neat EBMA (**4**) and BCA monomer (**5**). (**Right**): ~20,000 g/mol tPBCA using 25 mM EBMA (**1**), ~20,000 g/mol tPBCA using 5 mM EBMA (**2**), ~20,000 g/mol PBCA as comparison (**3**), neat EBMA (**4**) and BCA monomer (**5**).

Most importantly, the ^{13}C and ^{1}H spectra comparison of PBCA and tPBCA confirmed the integrity of the PBCA backbone structure after the addition of EBMA (Figure 8, ^{13}C data not shown). The EBMA vinyl signal in the ^{1}H spectrum was used to monitor the successful reaction between the EBMA and the polymer chain terminus. The different chemical shifts of the EBMA vinyl protons in the tPBCA spectra ($q^{\#}$) compared to the neat EBMA spectrum (q) confirmed that EBMA is chemically bonded to PBCA. The two proton signals of $q^{\#}$ were observed at a higher magnetic and lower magnetic field than the two protons of q, thus confirming the linkage between PBCA and EBMA. The observed band broadening of q[#] in all the tPBCA samples can again be assigned to the atactic structure of the polymer. However, the q signal from the neat EBMA was still detectable in the ^{13}C and ^{1}H spectra of tPBCA for the 2000 g/mol and 20,000 g/mol PBCA batches reacted with 25 mM EBMA (Figure 8). Thus, the neat EBMA vinyl group signal in the tPBCA spectra proved the presence of unreacted EBMA molecules. Apparently, 25 mM EBMA bears an excess of the end-capping agent for both MW batches in this trial. A concentration of 5 mM EBMA reacted with the 20,000 g/mol PBCA batch also exhibited the neat EBMA vinyl group signal (q) in both ^{13}C and ^{1}H spectra. The 5 mM EBMA reacted with the 2000 g/mol batch showed no neat EBMA vinyl group signal (q) in the ^{13}C and ^{1}H tPBCA spectra, indicating a complete reaction between EBMA and the polymer.

These results confirm that the required amount of terminating agent depends on the polymer MW and, consequently, on the number of polymer chain termini. In an equal amount of polymer, the higher MW batch has a lesser number of chain termini compared to the lower MW batch. Consequently, the lower the PBCA MW, the higher the required terminator amount to react with all chain termini. The optimal EBMA amount for the termination of 2000 g/mol PBCA is lower than 25 mM and the optimal EBMA amount for the termination of 20,000 g/mol is lower than 5 mM. However, further studies are necessary to identify the exact EBMA amount required for a specific MW so to avoid residual unreacted EBMA molecules. In addition, the ^{1}H spectra of all the tPBCA samples showed small amounts of residual BCA monomer (m), as described above for the PBCA polymers (Figure 8). The residual monomer probably originated from the NaOMe-induced monomer unzipping reaction during the DPRP prior to the addition of the end-capping agent EBMA. Based on the presence of the monomer in PBCA and tPBCA, a polymer stability study was conducted (see Section 3.2.3) to monitor the possible monomer release in PBCA and tPBCA, and to assess the stability, and thus the potential shelf-life of the polymers.

3.2.3. Polymer Stability in Solution

The stability of PBCA was monitored by measuring the increase of the monomer signal in the ^{1}H NMR spectra. A signal increase over time would indicate potential chemical structural changes such as chain deconstruction or degradation. Two batches of neat PBCA and corresponding tPBCA having a MW of 2000 g/mol and 20,000 g/mol, respectively, were stored in $CDCl_3$ solution for seven weeks. The ^{1}H NMR spectra were recorded once a week, focusing on the monomer signal at 7.058 ppm. The relative monomer content was estimated using the BHT signal at 6.98 ppm as a constant reference (constant = 1). Figure 9 depicts the constant BHT signal in all measured samples during the monomer release study of the 2000 g/mol PCBA. The monomer signal marked 'm' increased steadily over the seven week period.

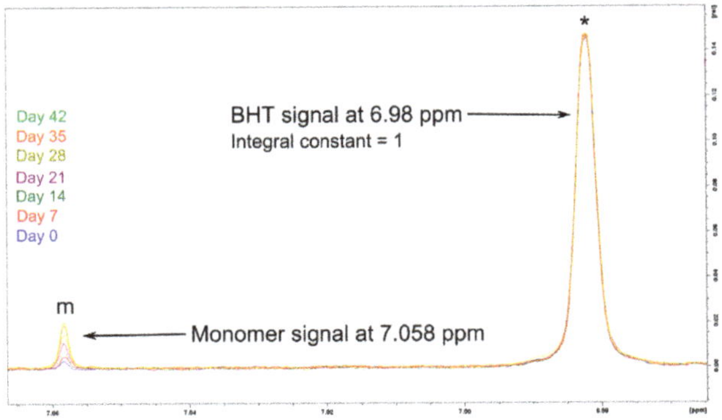

Figure 9. The ^{1}H NMR spectrum section of the monomer (m) and BHT (*) signal in 2000 g/mol PBCA stored for seven weeks as solution in $CDCl_3$. The BHT signal area (*) was constant during the storage, and consequently the integral value for BHT was assumed to be 1.

It can be assumed that the monomer residue originated from the NaOMe-derived monomer unzipping reaction (DPRP), resulting in the previously elucidated MW changes. However, the monomer release could also be a phenomenon induced by the living character of the PBCA chain progressing after the completion of the polymerization process. However, despite the living character of the PBCA as a polymer, it is essential that PBCA as a pharmaceutical excipient has a certain inherent chemical stability, especially to provide reasonable shelf-life. Thus, the end-capping of the chain termini using EBMA could potentially increase the shelf-life of PBCA as excipient by depriving PBCA of its living character and therewith preventing monomer release. The main observations made from the experimental results displayed in Figure 10 are:

1. The monomer release is indeed a progressive phenomenon for both PBCA batches, as the monomer amount clearly increased during storage;
2. The EBMA end-capping of the PBCA chains (tPBCA) inhibited the progression of the monomer release.

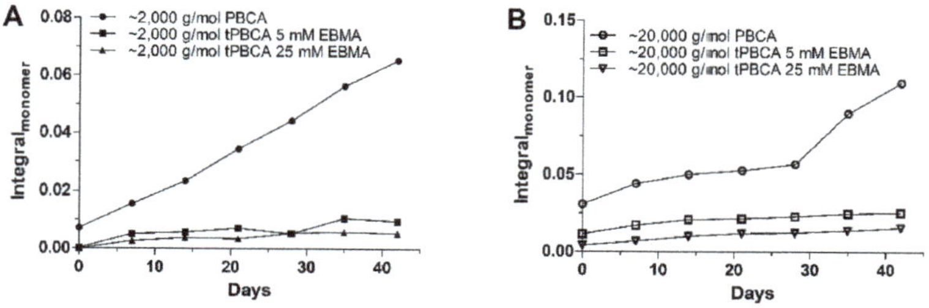

Figure 10. Monomer release monitored in ^{1}H NMR measurements at 7.058 ppm (m) for PBCA and tPBCA. tPBCA was terminated using two different EBMA concentrations, 5 mM and 25 mM. PBCA and tPBCA having a MW of ~2000 g/mol (**A**) and ~20,000 g/mol (**B**) were stored for seven weeks in $CDCl_3$ solution.

This proves that the PBCA chain is truly "alive", thus also confirming the hypothesis of the monomer unzipping reaction. It also proves that the termination with EBMA deprived the PBCA chain of its living character because PBCA lost its C-H acidity at the end chain, and therefore the formation of carbanion was not possible. Consequently, the PBCA chain restructuring process (DPRP) was prevented by the end-capping with EBMA, which effectively suppressed the monomer release.

4. Conclusions

The process design of the anionic living polymerization of BCA in THF by varying the amounts of TFA, NaOMe and BCA demonstrated that the MW of PBCA was controllable within a range of 1000 g/mol to 200,000 g/mol. It turned out that the key factor for MW control was the added amount of NaOMe. Characterization by APC™ proved to be a good tool to monitor the polymerization process, while the NMR measurements confirmed the expected chemical structure. Both techniques also confirmed the postulated depolymerization–repolymerization behavior of PBCA. Stability studies on neat PBCA and EBMA-end-capped PBCA in solution were conducted over the course of seven weeks. Neat PBCA revealed a larger growth of free monomer over time due to the depolymerization, whereas the monomer growth was suppressed in the end-capped PBCA. The ongoing depolymerization of the polymer chain confirmed the living character of PBCA. The chemical instabilities of the neat polymer would cause a shelf-life reduction, thus impacting the quality of PBCA as an excipient. The present work demonstrates that low and high MW PBCA can be synthesized and end-capped as a possible means to increase the chemical stability and, hence, the shelf-life of the polymer. Increasing the shelf-life can potentially enable the diverse drug delivery applications of this biodegradable polymer.

Author Contributions: Conceptualization, B.-L.K. and G.F.; synthesis, investigation and analysis, B.-L.K.; resources, S.O.K.; writing—original draft preparation, C.A.L. and B.-L.K.; writing—review and editing, C.A.L., B.-L.K. and S.O.K.; supervision, G.F.; project administration, G.F. and S.O.K. All authors have read and agreed to the published version of the manuscript.

Funding: This research received no external funding.

Institutional Review Board Statement: Not applicable.

Informed Consent Statement: Not applicable.

Data Availability Statement: Restrictions apply to the availability of these data. Data was obtained from AbbVie and are not available without the permission of AbbVie.

Acknowledgments: The authors thank Yeliz Tunc-Sarisozen and Cristian Cojocariu of Waters Corporation for reviewing of the manuscript, fruitful discussions and valuable input.

Conflicts of Interest: Benjamin-Luca Keller and Samuel Kyeremateng are employees of AbbVie and may own AbbVie stock. Claudia A. Lohmann is an employee of Waters Corporation and has no additional conflicts of interest to disclose. AbbVie sponsored and funded the study; contributed to the design; participated in the collection, analysis and interpretation of data, and in writing, reviewing and approval of the final publication.

References

1. Soppimath, K.S.; Aminabhavi, T.M.; Kulkarni, A.R.; Rudzinski, W.E. Biodegradable polymeric nanoparticles as drug delivery devices. *J. Control Release* **2001**, *70*, 1–20. [CrossRef]
2. Elzoghby, A.O.; Samy, W.M.; Elgindy, N.A. Albumin-based nanoparticles as potential controlled release drug delivery systems. *J. Control Release* **2012**, *157*, 68–82. [CrossRef] [PubMed]
3. Yadav, N.; Khatak, S.; Sara, U.V.S. Solid lipid nanoparticles—A review. *Int. J. Appl. Pharma* **2013**, *5*, 8–18.
4. Ling, D.; Hyeon, T. Chemical design of biocompatible iron oxide nanoparticles for medical applications. *Small* **2013**, *9*, 1450–1466. [CrossRef] [PubMed]
5. Boisselier, E.; Astruc, D. Gold nanoparticles in nanomedicine: Preparations, imaging, diagnostics, therapies and toxicity. *Chem. Soc. Rev.* **2009**, *38*, 1759–1782. [CrossRef] [PubMed]
6. Samad, A.; Sultana, Y.; Aqil, M. Liposomal drug delivery systems: An update review. *Curr. Drug Deliv.* **2007**, *4*, 297–305. [CrossRef] [PubMed]
7. Lacerda, L.; Bianco, A.; Prato, M.; Kostarelos, K. Carbon nanotubes as nanomedicines: From toxicology to pharmacology. *Adv. Drug Deliv. Rev.* **2006**, *58*, 1460–1470. [CrossRef] [PubMed]
8. Van, E.B.; Van den Mooter, G.; Augustijns, P. Top-down production of drug nanocrystals: Nanosuspension stabilization, miniaturization and transformation into solid products. *Int. J. Pharm.* **2008**, *364*, 64–75. [CrossRef]
9. Kumari, A.; Yadav, S.K.; Yadav, S.C. Biodegradable polymeric nanoparticles-based drug delivery systems. *Colloids Surf. B Biointerfaces* **2010**, *75*, 1–18. [CrossRef] [PubMed]
10. Couvreur, P.; Kante, B.; Roland, M.; Guiot, P.; Bauduin, P.; Speiser, P. Polycyanoacrylate nanocapsules as potential lysosomotropic carriers: Preparation, morphological and sorptive properties. *J. Pharm. Pharmacol.* **1979**, *5*, 331–332. [CrossRef]
11. Vauthier, C.; Dubernet, C.; Fattal, E.; Pinto-Alphandary, H.; Couvreur, P. Poly(alkylcyanoacrylates) as biodegradable materials for biomedical applications. *Adv. Drug Deliv. Rev.* **2003**, *55*, 519–548. [CrossRef]
12. Nicolas, J.; Couvreur, P. Synthesis of poly(alkyl cyanoacrylate)-based colloidal nanomedicines. *Wiley Interdiscip. Rev. Nanomed. Nanobiotechnol.* **2009**, *1*, 111–127. [CrossRef]
13. Gao, S.; Xu, Y.; Asghar, S.; Chen, M.; Zou, L.; Eltayeb, S.; Huo, M.; Ping, Q.; Xiao, Y. Polybutylcyanoacrylate nanocarriers as promising targeted drug delivery systems. *J. Drug Target.* **2015**, *23*, 481–496. [CrossRef] [PubMed]
14. *P010002*; Indermil™ Tissue Adhesive. U.S. Food and Drug Administration: Silver Spring, MD, USA, 2002.
15. Pifferi, G.; Santoro, P.; Pedrani, M. Quality and functionality of excipients. *Farmaco* **1999**, *54*, 1–14. [CrossRef]
16. Pifferi, G.; Restani, P. The safety of pharmaceutical excipients. *Farmaco* **2003**, *58*, 541–550. [CrossRef]
17. Eromosele, I.C.; Pepper, D.C. Anionic polymerization of butyl cyanoacrylate by tetrabutylammonium salts, Propagation rate constants. *Macromol. Chem. Phys.* **1989**, *190*, 3095–3103. [CrossRef]
18. Limouzin, C.; Caviggia, A.; Ganachaud, F.; Hemery, P. Anionic Polymerization of n-Butyl Cyanoacrylate in Emulsion and Miniemulsion. *Macromolecules* **2003**, *36*, 667–674. [CrossRef]
19. Almalik, A.; Donno, R.; Cadman, C.J.; Cellesi, F.; Day, P.J.; Tirelli, N. Hyaluronic acid-coated chitosan nanoparticles: Molecular weight-dependent effects on morphology and hyaluronic acid presentation. *J. Control Release* **2013**, *172*, 1142–1150. [CrossRef] [PubMed]
20. Mittal, G.; Sahana, D.K.; Bhardwaj, V.; Kumar, M.N.V.R. Estradiol loaded PLGA nanoparticles for oral administration: Effect of polymer molecular weight and copolymer composition on release behavior in vitro and in vivo. *J. Control Release* **2007**, *119*, 77–85. [CrossRef] [PubMed]

21. Gorner, T.; Gref, R.; Michenot, D.; Sommer, F.; Tran, M.N.; Dellacherie, E. Lidocaine-loaded biodegradable nanospheres. I. Optimization Of the drug incorporation into the polymer matrix. *J. Control Release* **1999**, *57*, 259–268. [CrossRef]
22. Ryan, B.; McCann, G. Novel sub-ceiling temperature rapid depolymerization-repolymerization reactions of cyanoacrylate polymers. *Macromol. Chem. Phys.* **1996**, *17*, 217–227. [CrossRef]
23. Pepper, D.C.; Ryan, B. Initiation Processes in Polymerization of Alkyl Cyanoacrylates by Tertiary Amines: Inhibition by Strong Acids. *Macromol. Chem. Phys.* **1983**, *184*, 383–394. [CrossRef]
24. Costa, G.; Cronin, D.C.; Pepper, D.C.; Loonan, C. Termination and transfer by acids in the pyridineinitiated. polymerization of butyl cyanoacrylate. *Euro Polym. J.* **1983**, *19*, 939–945. [CrossRef]
25. Odian, G. *Principles of Polymerization*, 4th ed.; John Wiley & Sons, Inc.: Hoboken, NJ, USA, 2004; pp. 412–436. [CrossRef]
26. Donnelly, E.F.; Johnston, D.S.; Pepper, D.C. Ionic and zwitterionic polymerization of *n*-alkyl 2-cyanoacrylates. *Polym. Lett. Ed.* **1977**, *15*, 399–405. [CrossRef]
27. Canale, A.J.; Goode, W.E.; Kinsinger, J.B.; Panchak, J.R.; Kelso, R.L.; Graham, R.K. Methyl α-cyanoacrylate I Free-radical homopolymerization. *J. Appl. Polym. Sci.* **1960**, *4*, 231–236. [CrossRef]
28. Han, M.G.; Kim, S.; Liu, S.X. Synthesis and degradation behavior of poly(ethyl cyanoacrylate). *Polym. Degrad. Stab.* **2008**, *93*, 1243–1251. [CrossRef]
29. Miele, E.; Spinelli, G.P.; Miele, E.; Tomao, F.; Tomao, S. Albumin-bound formulation of paclitaxel (Abraxane ABI-007) in the treatment of breast cancer. *Int. J. Nanomed.* **2009**, *4*, 99–105. [CrossRef]
30. Robello, D.R.; Eldridge, T.D.; Swanson, M.T. Degradation and Stabilization of Polycyanoacrylates. *J. Polym. Sci.* **1999**, *37*, 4570–4581. [CrossRef]
31. Leonard, F.; Kulkarni, R.K.; Brandes, J.N.; Cameron, J. Synthesis and Degradation of Poly(alkyl-cyanoacrylates. *J. Appl. Polym. Sci.* **1966**, *10*, 259–272. [CrossRef]
32. Pirker, S.; Kruse, J.; Noe, C.; Langer, K.; Zimmer, A.; Kreuter, J. Characterization of polybutyleyanoacrylate nanoparticles. Part II: Determination of polymer content by NMR analysis. *Int. J. Pharm.* **1995**, *128*, 189–195. [CrossRef]
33. Wohlgemuth, M.; Machtle, W.; Mayer, C. Improved preparation and physical studies of polybutylcyanoacrylate nanocapsules. *J. Microencapsul.* **2000**, *17*, 437–448. [CrossRef]
34. Mulik, R.; Mahadik, K.; Paradkar, A. Development of curcuminoids loaded poly(butyl)cyanoacrylate nanoparticles: Physicochemical characterization and stability study. *Eur. J. Pharm. Sci.* **2009**, *37*, 395–404. [CrossRef]
35. Mayer, C. NMR Studies of Nanoparticles. *Annu. Rep. NMR Spectrosc.* **2005**, *55*, 258. [CrossRef]
36. Szanka, I.; Szanka, A.; Kennedy, J. Robbery Wound Closure Adhesives. II. Initiators for and Initiation of 2-Octyl Cyanoacrylate Polymerization. *J. Polym. Sci.* **2015**, *53*, 1652–1659. [CrossRef]
37. Kohsaka, Y.; Ishihara., S.; Kitayama, T. Termination of Living Anionic Polymerization of Butyl Acrylate with alpha-(Chloromethyl)acrylate for End-Functionalization and Application to Evaluation of Monomer Reactivity. *Macromol. Chem. Phys.* **2015**, *216*, 1534–1539. [CrossRef]

polymers

MDPI

Article

Concentration Effect over Thermoresponse Derived from Organometallic Compounds of Functionalized Poly(*N*-isopropylacrylamide-*co*-dopamine Methacrylamide)

María Moral-Zamorano [1], Isabel Quijada-Garrido [2], Verónica San-Miguel [1], Berna Serrano [1], Juan Baselga [1], Saud Hashmi [3], Florian J. Stadler [4,*] and Alberto García-Peñas [1,*]

1 Departamento de Ciencia e Ingeniería de Materiales e Ingeniería Química, IAAB, Universidad Carlos III de Madrid, Avda. de la Universidad, 30, 28911 Leganés, Madrid, Spain; mamoralz@ing.uc3m.es (M.M.-Z.); vmiguel@ing.uc3m.es (V.S.-M.); berna@ing.uc3m.es (B.S.); jbaselga@ing.uc3m.es (J.B.)
2 Instituto de Ciencia y Tecnología de Polímeros ICTP-CSIC, Juan de la Cierva 3, 28006 Madrid, Spain; iquijada@ictp.csic.es
3 Department of Polymer and Petrochemical Engineering, NED University of Engineering and Technology, University Road, Karachi 75270, Pakistan; saudhashmi@cloud.neduet.edu.pk
4 College of Materials Science and Engineering, Shenzhen Key Laboratory of Polymer Science and Technology, Guangdong Research Center for Interfacial Engineering of Functional Materials, Nanshan District Key Laboratory for Biopolymers and Safety Evaluation, Shenzhen University, Shenzhen 518055, China
* Correspondence: fjstadler@szu.edu.cn (F.J.S.); alberto.garcia.penas@uc3m.es (A.G.-P.)

Citation: Moral-Zamorano, M.; Quijada-Garrido, I.; San-Miguel, V.; Serrano, B.; Baselga, J.; Hashmi, S.; Stadler, F.J.; García-Peñas, A. Concentration Effect over Thermoresponse Derived from Organometallic Compounds of Functionalized Poly(*N*-isopropylacrylamide-*co*-dopamine Methacrylamide). *Polymers* **2021**, *13*, 3921. https://doi.org/10.3390/polym13223921

Academic Editor: Vijay Kumar Thakur

Received: 12 October 2021
Accepted: 10 November 2021
Published: 12 November 2021

Publisher's Note: MDPI stays neutral with regard to jurisdictional claims in published maps and institutional affiliations.

Abstract: The functionalization of smart polymers is opening a new perspective in catalysis, drug carriers and biosensors, due to the fact that they can modulate the response regarding conventional devices. This smart response could be affected by the presence of organometallic complexes in terms of interactions which could affect the physical chemical properties. In this sense, the thermoresponsive behavior of copolymers based on *N*-isopropylacrylamide (NIPAM) could be affected due to the presence of hydrophobic groups and concentration effect. In this work, the functionalization of a copolymer based on NIPAM and dopamine methacrylamide with different amounts of bis(cyclopentadienyl)titanium (IV) dichloride was carried out. The resulting materials were characterized, showing a clear idea about the mechanism of functionalization through FTIR spectroscopy. The thermoresponsive behavior was also studied for various polymeric solutions in water by UV–vis spectroscopy and calorimetry. The hydrophobic interactions promoted by the organometallic complex could affect the transition associated with the lower critical solution temperature (LCST), specifically, the segments composed by pure NIPAM. That fact would explain the reduction of the width of the LCST-transition, contrary to what could be expected. In addition, the hydrophobicity was tested by the contact angle and also DNA interactions.

Keywords: thermoresponsive polymers; hydrophobic transitions; lower critical solution temperature; functionalized materials; contact angle

1. Introduction

The new strategies for sustainable progress developed by some governments are affecting industry, which needs to look for alternatives to traditional methods of production and also for new eco-friendly materials. Those changes must maintain companies' profitability and provide other benefits such as improving the efficiency of industrial processes.

The generation of smart materials is a valuable way to respond to a part of industrial demand as they can improve the efficiency of some processes, reduce the consumption of energy, and are biodegradable in some cases where biodegradable monomers are involved as well [1–15]. Many efforts focus on the preparation and characterization of these materials, but the composites and functionalized polymers are growing as those can provide specific and modulated properties. In this sense, the preparation of smart polymeric matrices with inorganic compounds can provide additional functions to the pure polymers used in many applications [16–18]. Nevertheless, the problems associated with the lability of the metal–carbon bonds partially stopped this progress, but new regulation and advances in macromolecular chemistry have renewed interest in these structures [19,20].

There are some recent advances reported where smart polymers are involved in the preparation of hybrid materials for heterogeneous catalysis, which offer some advantages in terms of easy separations and recyclability [21–23]. However, applications can be extended to other fields as biomedical applications [8,24–28] or nanotechnology [29].

The final properties of the new materials will depend on the molecular details associated with the polymer and the organometallic complex, but also the relation between both will play an important role [30]. For example, a sufficiently high content of an organometallic complex could affect the physicochemical properties of the polymers. On the other hand, the functionalization of the smart polymers with organometallic complexes can be limited by different reasons like the numerous steps involved in the procedures, hazardous solvents, or the limitations for incorporating high amounts of an organometallic complex into the polymeric matrices, among other factors.

The use of comonomers into the polymeric chains, as catechol groups, can provide other functionalities or reversibility. Recently, our research group proposed the novel and facile functionalization of the –OH groups of poly(*N*-isopropylacrylamide-*co*-dopamine methacrylamide), where we observed some particularities associated with the presence of the organometallic complexes [31]. Those copolymers were functionalized with an organotin (IV) compound and bis(cyclopentadienyl)titanium (IV) dichloride showing the hydrophobic interactions promoted by the organometallic complexes over the surrounding polymeric chains. Nevertheless, these preliminary works presented and analyzed functionalized materials with a minimum amount of metallodrugs as we wanted to preserve the original structure of the materials and also keep the thermoresponsive nature of the polymers [8,31]. In this sense, there are many questions that need to be addressed about the presence of organometallic complexes in smart polymeric chains, such as: How could affect the presence of the organometallic complex to the lower critical solution temperature (LCST) transition? What kind of interactions can be produced between polymeric chains and their hydrophobic units composed by the comonomer and organometallic complexes? Could the concentration of functionalized polymers in water change the LCST behavior due to the presence of surrounding polymeric chains?

This work tries to take a step towards answering these by working with different amounts of organometallic complexes over the same smart polymer, analyzing the changes promoted into the matrix and the thermoresponsive behavior of the corresponding polymeric solutions prepared with different concentrations in water. For that purpose, a copolymer based on *N*-isopropylacrylamide and dopamine methacrylamide was prepared and divided into different fractions. Each fraction of the copolymer was functionalized with a specific amount of bis(cyclopentadienyl)titanium (IV) dichloride, obtaining a wide range of different samples with different amounts of the organometallic complex. A complete characterization of the molecular features was carried out through proton nuclear magnetic resonance, Fourier-transform infrared spectroscopy, X-ray fluorescence, and X-ray diffraction. The thermoresponsive behavior was studied by differential scanning calorimetry and ultraviolet–visible (UV–vis) spectroscopy for both polymeric aqueous solutions (1 and 4 wt.%). The hydrophobicity was analyzed by the contact angle, and also, the DNA interactions were tested for the new materials.

2. Materials and Methods

2.1. Materials

N-isopropylacrylamide (NIPAM) (98%, Biosynth Carbosynth, Compton, UK) was purified by recrystallization using toluene (99%, Alfa Aesar, Haverhill, MA, USA) and hexane (96%, Scharlau, Sentmenat, Spain). Tetrahydrofuran (99.8%, Scharlau, Barcelona, Spain) and *N,N*-dimethylformamide (DMF) (99.8%, Alfa Aesar, Kandel, Germany) were dried before use. The azobisisobutyronitrile (AIBN) (98%, Sigma Aldrich, Saint-Quentin-Fallavier, France) was recrystallized from methanol (99.99%, Quimipur, Campo Real, Spain).

The preparation of the dopamine methacrylamide (DMA) involved the use of sodium borate (98%, Panreac, Barcelona, Spain), sodium bicarbonate (99%, Fluka, Hamburg, Germany), methacrylate anhydride (94%, Sigma Aldrich, Hamburg, Germany), 3,4-dihydroxyphenethylamine hydrochloride (98.5%, Fluka, Germany), sodium hydroxide (97%, Sigma Aldrich, Hamburg, Germany), ethyl acetate (99.91%, Quimipur, Campo Real, Spain), hydrochloric acid (37%, Sigma Aldrich, Hamburg, Germany), water (Quality Level: 200, Sigma Aldrich, Buchs, Switzerland) and magnesium sulfate (97%, Panreac, Europe Union).

The resulting polymer was precipitated in diethyl ether (99.7%, Quimipur, Campo Real, Spain). The functionalization was carried out using triethylamine (NEt3) (99%, Sigma Aldrich, Overijse, Belgium), and bis(cyclopentadienyl)titanium (IV) dichloride (also called titanocene dichloride, Cp_2TiCl_2) (97%, Sigma Aldrich, Moscow, Russia). The DNA interactions were performed with Buffer (Hyclone Products Cytiva, Amersham, UK) and salmon DNA ($\leq$5% protein, Sigma Aldrich, Tokyo, Japan).

2.2. Synthesis of Dopamine Methacrylamide

Dopamine methacrylamide (DMA) was prepared following the procedure previously published in the literature [32]. Subsequently, the DMA was purified by precipitation in hexane.

2.3. Preparation of Poly(N-Isopropylacrylamide-Co-Dopamine Methacrylamide) Copolymer

The copolymer based on *N*-isopropylacrylamide and dopamine methacrylamide was synthesized using a Schlenk tube under a nitrogen atmosphere. For that purpose, the *N*-isopropylacrylamide (0.0663 mol), dopamine methacrylamide (0.0046 mol), and AIBN initiator (7×10^{-5} mol) were added inside the Schlenk tube. Then, 10 mL DMF was injected in the tube, keeping the inert conditions. The Schlenk tube was located inside of a thermostatic bath at 70 °C where the reaction took place for 48 h. After, the reaction was stopped using nitrogen liquid, the copolymer was precipitated in diethyl ether.

The final polymer was washed and purified several times through precipitation, dried under vacuum and, finally, it was stored at room temperature. The details about the synthetic procedure and reaction schemes can be found in previous works reported by the authors [31,33]. The yield of the reaction obtained was around 68%.

2.4. Functionalization Process with Bis(cyclopentadienyl)titanium (IV) Dichloride

The copolymer was divided into four parts which were functionalized with different amounts of bis(cyclopentadienyl)titanium (IV) dichloride. For that purpose, each part of the copolymer was independently dissolved in Schlenk tubes using DMF as the solvent and under nitrogen atmosphere. A solution of triethylamine (0.0359 mol·L^{-1}) was prepared in advance, and specific amounts were introduced inside the Schlenk tubes where, previously, polymeric solutions were prepared, as shown in Table 1. Then, an organometallic complex solution (0.0161 mol·L^{-1}) was injected in the previous tubes following the amounts described in Table 1, for obtaining materials with different concentrations of organometallic complex in a wide range of compositions. The reaction was carried out at room temperature for 36 h, and subsequently, the polymer was precipitated in diethyl ether, washed, and filtered under inert conditions. Finally, the samples were dried under vacuum at room temperature.

From now on, the pristine copolymer is denoted as cDMA, adding f for functionalized samples (cDMAf) which bis(cyclopentadienyl)titanium (IV) dichloride (Cp_2TiCl_2) content varies and increases (indicated from 1 to 4) as collected in Table 1.

Table 1. Functionalization conditions for copolymers functionalized with the organometallic complex (cDMAf1-4).

Sample	DMF [mL]	NEt3 [mol]	Cp_2TiCl_2/cDMA [mole Ratio]
cDMA	10	0	0
cDMAf1	10	2.03×10^{-6}	2.43×10^{-4}
cDMAf2	10	4.06×10^{-6}	4.87×10^{-4}
cDMAf3	10	2.03×10^{-5}	2.43×10^{-3}
cDMAf4	10	4.06×10^{-5}	4.87×10^{-3}

2.5. Preparation of Polymeric Solutions

The phase transition temperature associated with the LCST and the contact angle was studied for the pure copolymer and functionalized materials. For that purpose, two polymeric concentrations in water (1 and 4 wt.%) were prepared, depending on the technique and its resolution and for checking the differences associated with the interactions between polymeric chains in water. The solutions in water were kept in the refrigerator for homogenization of the samples for 24 h before measurements.

All measurements were carried out at least in triplicate, which guarantees the consistency of these results.

2.6. Gel Permeation Chromatography

Molecular weight distributions and polydispersity (M_w/M_n) were determined by a Waters SEC system (Milford, MA, USA) equipped with a Waters 1515 Isocratic HPLC Pump and a Waters 2414 refractive index detector, using DMF with LiBr 0.1 wt.% as the mobile phase at a flow rate of 0.7 mL·min^{-1} and 70 °C using polystyrene standards for the calibration.

2.7. X-ray Diffraction

The measurements were carried out in a Philips X'Pert diffractometer provided with a PW3011/10 detector (Eindhoven, The Netherlands). The diffraction scans were collected between 5 and 80° with a 2θ step of 0.05° and 2.5 s per step.

2.8. Fourier Transform Infrared (FTIR) Spectroscopy

FTIR spectra were recorded by a Perkin-Elmer Spectrum Two FTIR-Spectrometer (Waltham, MA, USA) fitted with attenuated total reflectance (ATR). The samples were placed in direct contact with the diamond crystal without additional preparation. Measurements were collected from 4000 to 400 cm^{-1} at a resolution of 4 cm^{-1} with 32 scans per spectrum.

2.9. Nuclear Magnetic Resonance (^1H NMR) Spectroscopy

^{1}H-NMR spectra were recorded in deuterated DMSO at 400 MHz on a Bruker Avance III-HD 400 spectrometer (Billerica, MA, USA).

2.10. X-ray Fluorescence Spectrometry

The content of organometallics into polymeric chains was estimated by X-ray fluorescence spectrometer (Ametek Materials Analysis Division Spectro Xepos, Devon-Berwyn, USA), to detect the Ti content of the samples.

2.11. Ultraviolet–Visible (UV-Vis) Spectroscopy

Cloud point measurements were carried out in a Cary 3 BIO-Varian UV–Visible spectrophotometer (Palo Alto, CA, USA) equipped with a Peltier temperature control device. The temperature was raised from 12 to 45 °C at a heating rate of 1 °C·min^{-1}.

The estimation of the cloud points and phase transition temperatures were carried out as previously reported [34].

2.12. Differential Scanning Calorimetry

The polymeric solutions (around 15 mg) were placed in aluminum pans for studying the phase transition temperature associated with LCST by differential scanning calorimeter (UC3M: Mettler Toledo DSC822e, L'Hospitalet de Llobregat, Spain) equipped with a cooling system. All the tests were carried out between 0 and 45 °C, using a heating rate of 5 °C·min^{-1}. The calorimetric curves were normalized using another experiment for pure water.

2.13. Contact Angle

Contact angles were measured using a Dataphysics Contact Angle System OCA Camera: Teli CCD Camera for all the samples at different temperatures (below and above LCST). Measurements were carried out in a range from 15 to 40 °C every 5 °C. Dry polymer films were prepared by drop-casting, placing 80 µL of an aqueous solution (2 wt.%) on a glass surface with a Teflon frame for keeping a fixed diameter of 1 cm.

Finally, 2 µL water droplets were dispensed onto the dried surface of the polymeric film. The average contact angle value was obtained from at least three measurements for each copolymer.

2.14. DNA Interactions

First, each functionalized copolymer (25 mg) was mixed with ethanol (25 mL) and phosphate buffer solution (25 mL) and was stored in a refrigerator (4 °C). On the other hand, several DNA solutions (0.02, 0.03, 0.04, 0.05, 0.06, and 0.08 mg/mL) were prepared. Then, polymeric solutions (3 mL) were mixed with each DNA-solutions (3 mL), and shaken at 35 °C (30 min) before measurements.

The analysis was carried out using an UV–vis spectrometer using a scanning wavelength from 800 nm to 200 nm at 35 °C. The samples were compared with pure DNA the absorbance of which is between 260 and 280 nm.

3. Results and Discussion

The molecular characterization was carried out for the pure copolymers and their corresponding functionalized samples. It was expected from synthetic and preparation routes to obtain different degrees of organometallic functionalization in a wide range of compositions. Therefore, the pure copolymer structure could suffer molecular changes as organometallic moieties are incorporated, varying the final spectrum of properties of functionalized materials.

The average molecular weight (M_n) and polydispersity of the pure copolymer, analyzed by gel permeation chromatography, were 42,000 g·mol^{-1} and 2.9, respectively. The relatively high polydispersity can be clearly explained by the free radical polymerization. The use of RAFT agents or catalysts was avoided as those could negatively affect biocompatibility. Also, these could disrupt the effect of the content of the organometallic complex over the molecular features and structure of the polymer, and consequently, on the thermoresponsive behavior and hydrophobicity of the samples. The comonomer content was estimated by ^{1}H NMR following the procedure described in the literature, obtaining a DMA content of 5.3 mol % [35].

On the other hand, the DMA content of polymeric chains was estimated by comparing the protons of the benzene ring and the protons of the -CH- of the side chain of NIPAM obtained ^{1}H NMR, as reported previously [31,33].

The molecular features of the resulting modified copolymers with Cp_2TiCl_2, and possible changes on the structures, were analyzed by FTIR spectroscopy, a powerful technique to elucidate interactions in organic-inorganic materials. FTIR spectra corresponding to PNIPAM, pristine cDMA, and copolymers functionalized with organometallic complexes are shown in Figure 1A. Differences between materials must be exclusively associated with the presence of organometallic complexes, as all the samples were prepared using the same copolymer composition. Typical vibration bands attributed to NIPAM and DMA appear in the spectra, at 3400–3300 cm^{-1} the O–H and N–H stretching, at 2970, 2930, and 2872 cm^{-1} the aliphatic C–H stretching, the strong bands (1640 and 1533 cm^{-1}) corresponding to the C=O stretching of the amide (amide I) and N–H bending mixed with C–N stretching (amide II), at 1386 and 1366 cm^{-1} the typical doublet, corresponding to the deformation vibration of the isopropyl group in NIPAM. Significant changes are observed between 1000 and 1300 cm^{-1}, specifically associated with the vibration bands observed at 1065 and 1253 cm^{-1}, which are identified with ν(C–O) bonds related to catechol group. In this sense, the band placed at 1253 cm^{-1} is related to C–OH bonds, which intensity probably changes due to the functionalization of organometallic complexes through those -OH groups. Our previous works suggested this route of functionalization [8,31], but this is the first proof that could confirm that organometallic moieties are linked to the copolymer by the -OH groups.

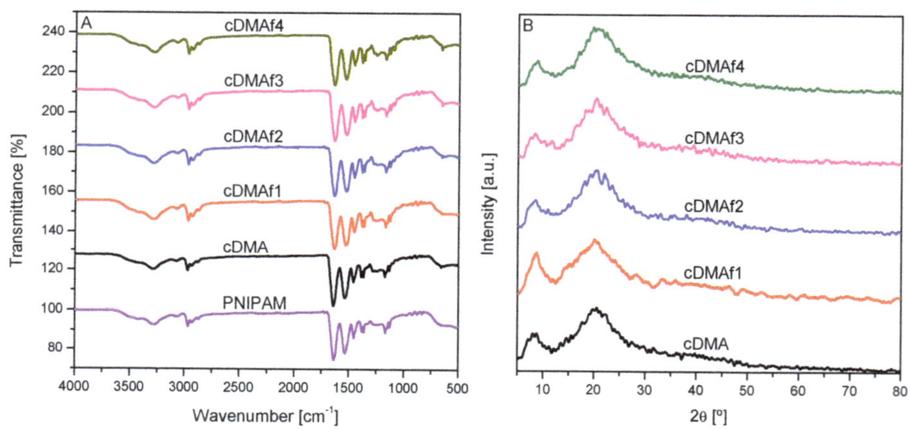

Figure 1. Microstructural features of the resulting materials characterized by Fourier transform infrared (FTIR) spectroscopy (**A**) and X-ray powder diffraction (**B**).

The FTIR spectroscopy can provide an idea about the degree of functionalization, which can be obtained through the band situated at 665 cm^{-1}. That band is identified with δ(=C–H) bonds and is associated with the aromatic rings of Cp_2TiCl_2. The change of intensity could respond to the amount of organometallic complex incorporated into the polymeric chains.

X-ray diffraction patterns show exclusively two broad peaks, which are associated with the poly(*N*-isopropyl acrylamide) diffraction pattern (Figure 1B). [15,36,37] Differences between pristine copolymer and functionalized samples were not detected, even for the sample with the highest amount of Cp_2TiCl_2, suggesting that the organometallic complex was homogeneously distributed into the polymeric chains and did not lead to phase separation or crystallization; besides the concentration of organometallic complex does not affect the shape of the diffractogram probably because amounts are not high enough or the low resolution of the technique.

The functionalization of the resulting polymers was also analyzed by ^1H NMR, as Figure 2A displays. Significant changes cannot be detected in the NMR spectra as the organometallic compound content is very low compared to the total content of DMA inserted along the polymeric chains. Another important problem is related to the overlapping between proton signals corresponding to Cp_2TiCl_2 and those of the pure copolymer, as can be deduced from their respective spectra. Due to the interaction of DMA with the organometallic compound, we would expect to observe specific changes in the peaks placed at 8.6 and 8.75 ppm, associated with the protons of -OH groups. [31] Nevertheless, the low content of the organometallic complex with respect to the DMA units inserted along the polymeric chains could explain, together with the overlapping with the main chain peaks, the lack of changes in the spectra. The similarity between the different spectra seems to indicate that structural changes are avoided.

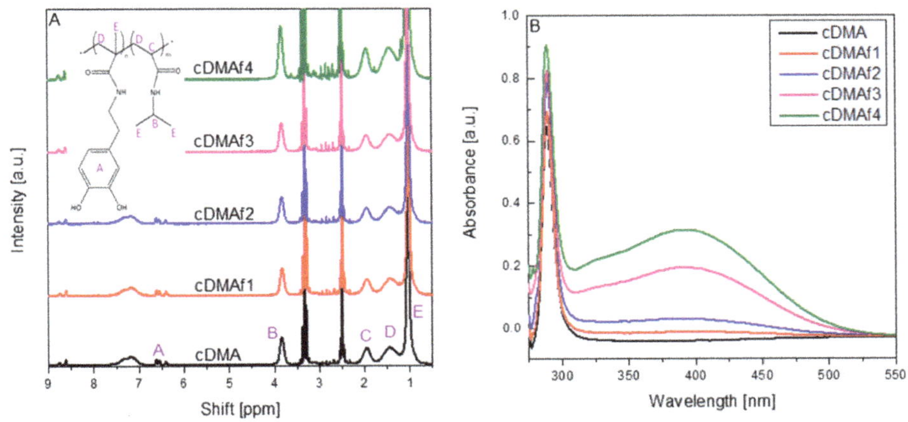

Figure 2. Molecular characterization for all the samples carried out by nuclear magnetic resonance (^{1}H NMR) in DMSO-d6 (**A**), and ultraviolet–visible (UV–vis) spectroscopy (**B**).

The functionalization of the samples was studied by UV–vis spectroscopy in aqueous solutions at ambient temperature as Figure 2B shows. The pure copolymer shows exclusively the absorbance of the C=O bonds, which are part of the NIPAM and DMA structure, but a second peak is identified for the functionalized samples. That second absorbance peak could be associated with the cyclopentadienyl groups and Ti-O bonds [38]. Furthermore, the concentration of organometallic compounds along the polymeric chains seems to play an important role, too, as the intensity of the peak increases as Cp_2TiCl_2 rises.

The elemental analysis was carried out by X-ray fluorescence for all the functionalized materials (Table 2). The percentage of Ti increases as was expected during the experimental work, showing the successful incorporation of organometallic compounds along the polymeric chains. Another important piece of information derived from elemental analysis is the relationship between Ti and Cl, which shows a value higher than 1 for cDMAf3 and cDMAf4. There are two possible routes for functionalization, i.e., this one may be carried out by the reaction of a single or both chlorine ligands through both OH-groups attached to the DMA (Scheme 1). The ratio between Ti and Cl seems to explain that a great part of the organometallic moieties is attached by a single bond to the polymeric structure as they seem to keep another chlorine (Scheme 1B). Nevertheless, these results are not conclusive, and the error associated with the measurements should be considered because it could play an important role.

Table 2. Percentage of Ti and Cl incorporated into the copolymers determined by X-ray fluorescence (wt.%).

Sample	Cl [%]	Error	Ti [%]	Error
cDMAf1	0.0029	0.00001	0.00049	0.00001
cDMAf2	0.0590	0.00010	0.02013	0.00017
cDMAf3	0.3543	0.00030	0.41120	0.00090
cDMAf4	0.5165	0.00040	0.67320	0.00120

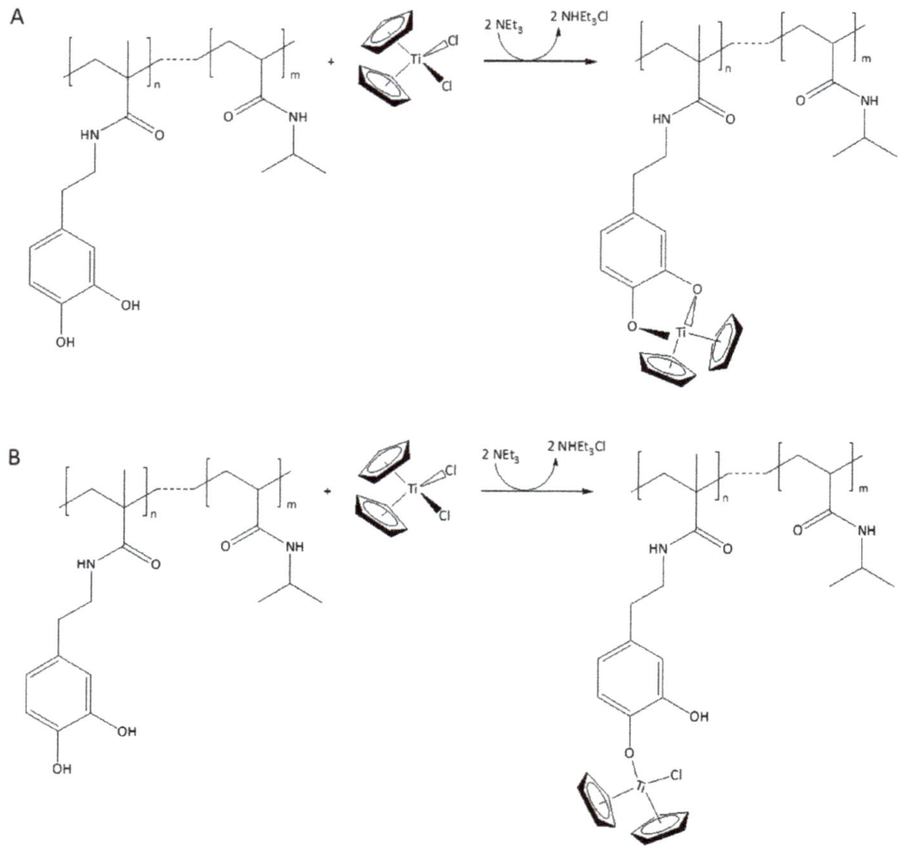

Scheme 1. Possible routes of functionalization between copolymer and organometallic complex through both OH groups (**A**) or a single OH (**B**).

The thermoresponsive properties in aqueous solutions were studied for all the functionalized samples and the pure copolymer by UV–vis spectroscopy. Figure 3 displays the phase transitions associated with the LCST for polymeric solutions (4 wt.%), showing the transitions pretty close to each other. Nevertheless, significant differences can be observed from the shape of the curves. In general, a higher number of hydrophobic groups promotes broader transitions [33]. The explanation related to these broad hydrophobic transitions was reported before for similar samples, but the amount of comonomer was varied, keeping the same amount of the organometallic complexes [31]. In that case, rheology showed that the phase transition temperature associated with LCST comprises two partially overlapped transitions. One of them is related to NIPAM sequences free of DMA placed at the highest temperature, and another one is situated at a lower temperature associated with the segments of the chains with the presence of DMA. Higher content of DMA into the polymeric chains will induce broader transitions as the lowest transition temperature associated with segments with comonomer will be placed at a lower temperature, increasing its distance with the phase transition temperature of pure NIPAM sequences.

This explanation cannot apply to the content of organometallic complexes for these samples with the same composition of DMA as it can be observed the opposite effect, i.e., transitions are wider as the content of hydrophobic organometallic complexes rises. It could be feasible to think about similar effects as both comonomer and organometallic complexes present a hydrophobic nature. Nevertheless, the situation is somewhat different as expected and, consequently, some parameters such as polymer concentration in water, interactions between surrounding chains, organometallic complex amount, and position of the organometallic moieties into the polymeric chains could play an important role.

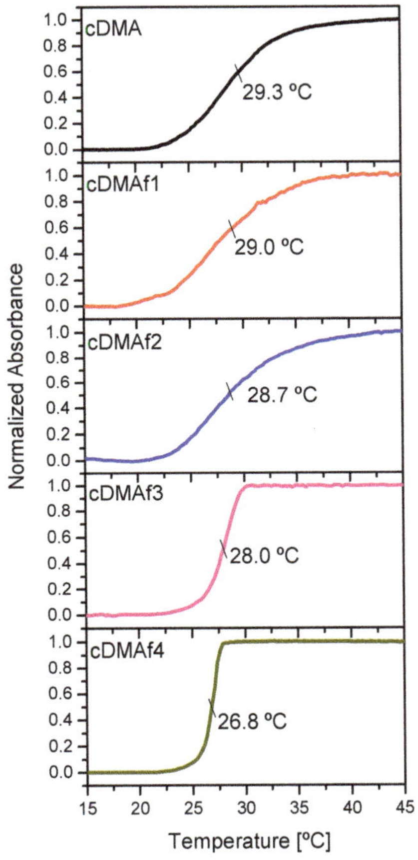

Figure 3. Phase transition temperatures, analyzed by UV–vis spectroscopy for all polymeric solutions (4 wt.%).

The polymer concentration in water could be high enough to allow interactions between polymeric chains which could be restricted by repulsive forces between DMA units and organometallic moieties. Nevertheless, a hydrophobic influence could be exerted by the organometallic complex over sections of surrounding polymeric chains composed by pure NIPAM. This influence could reduce the phase transition temperatures of LCST associated with these pure NIPAM sequences. Scheme 2 shows a clear idea about the differences between both effects. In the left part of the figure, the comonomer effect clearly will reduce the low phase transition temperature associated with the NIPAM + DMA sequences as comonomer content rises. The comonomer content will not affect the LCST of the NIPAM sequences, as they could not reach the main backbone due to repulsive forces between surrounding polymeric chains. This will lead to an increase in the width of the transition as the lowest LCST (NIPAM + DMA sequences) is reduced and the distance with the highest LCST (NIPAM sequences) increases.

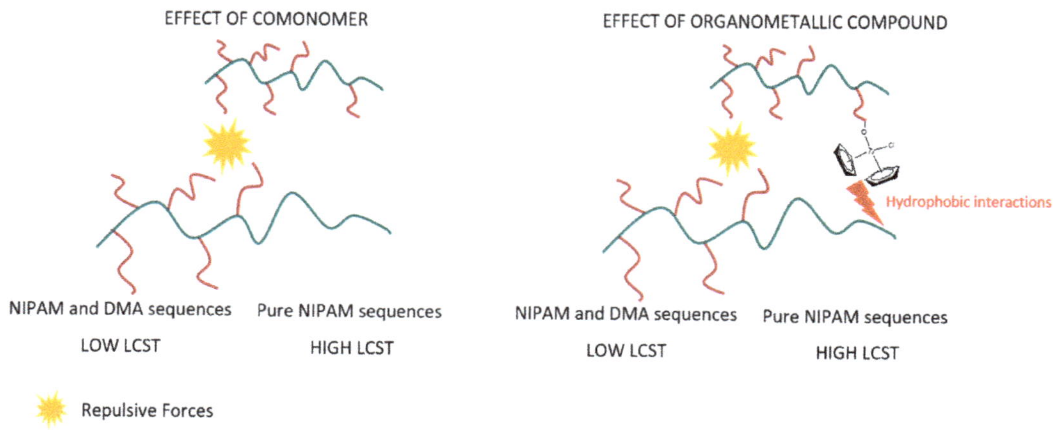

Scheme 2. Effect of hydrophobic groups over surrounding polymeric chains in solution.

The situation is rather different when the content of the organometallic complex rises to keep the same comonomer composition (right part of Scheme 2). The lowest LCST (NIPAM + DMA sequences) would be affected by organometallic compounds due to the repulsive forces between organometallic moieties and DMA units of surrounding polymeric chains that could protect part of the hydrogen bonds, increasing the lowest LCST. Those normally could be affected by the units of DMA of surrounding polymeric chains reducing the LCST. The content of organometallic complexes is too low compared to the comonomer content, but enough to partially reduce those interactions as organometallic compounds are too far from the main backbone. On the other hand, these organometallic compounds could probably affect pure NIPAM sequences of surrounding polymeric chains, decreasing the highest LCST, consequently inducing shorter hydrophobic transitions.

The cloud points were estimated for all the curves, and those decrease as the content of organometallic compounds rises, as expected. The hydrophobic groups could reduce the LCST, but this assumption needs to be studied in depth as molecular features of the polymeric chain in terms of composition, position, and distributions play an important role in the LCST mechanism.

Similar experiments were carried out by differential scanning calorimetry using the same polymer concentration in water but with higher heating rates because temperature control in a calorimeter is possible up to significantly higher rates than in optical cuvettes. These curves are displayed in Figure 4A, and it can be observed that the width of the hydrophobic transitions decreased as the content of organometallic complexes rose. The width of the transitions follows the trend observed by turbidimetry in Figure 3 but the maximum of the transitions does not look like to follow a specific tendency with the presence and the content of Cp_2TiCl_2. Turbidimetry showed that the cloud temperatures decreased as the organometallic complexes increased, but in this case, that behavior is not unambiguous. The differences between techniques could explain these changes, as UV–vis spectroscopy shows the reduction of the transmittance as increasing the phase separation structures, but calorimetry is very sensitive to the detection of the coil to globule transitions [34,39]. Many factors could explain that a clear tendency of the maximum of DSC is not exhibited, as end-group effect or the polymer–solvent interactions, which could be affected by the presence of the organometallic compound [40–43]. The reproducibility of the experiments confirms the results, and consequently, different phenomena should be responsible for this kind of curve.

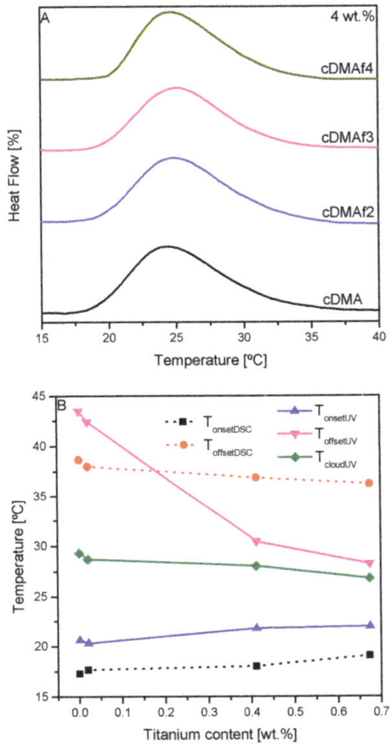

Figure 4. Phase transition temperatures associated with lower critical solution temperature (LCST), analyzed by calorimetry (**A**), and a comparison of calorimetric and UV results (**B**) for all polymeric solutions (4 wt.%).

Figure 4B shows the LCST transitions analyzed from curves of UV–vis spectroscopy and calorimetry. For that purpose, the onset and offset temperatures (T_{onset} and T_{offset}) were defined where the transitions or LCST-ranges start and finish, respectively. Also, cloud points were included, which were estimated at 50% of the UV–vis transition [33]. The transitions clearly diminish their width as the content of the organometallic complex increases, i.e., onset and offset temperatures are closer to each other. Another important fact can be deduced from the cloud temperatures, whose tendency is defined by the content of the organometallic complex but clearly defined in the middle of the onset and offset temperatures estimated from DSC. In general, the UV transitions are shorter than those determined by calorimetry, as could be expected due to calorimetry being more sensitive [34].

$T_{offsetUV}$ seems clearly affected in a higher proportion than $T_{onsetDSC}$ by the content of the organometallic complex due to $T_{offsetUV}$ being closer to the cloud points. $T_{offsetDSC}$ and $T_{onsetDSC}$ are defined proportionally at similar distances of the cloud points.

A small protection of a small part of the hydrogen bonds derived from the repulsive forces exerted between the organometallic moieties and the DMA of surrounding polymeric chains could explain the increase of the $T_{onsetUV}$ and $T_{onsetDSC}$, as explained above. On the other hand, the hydrophobic interactions promoted by the organometallic compounds over the pure NIPAM-sequences of the polymeric chains could also explain the decrease of the $T_{offsetUV}$ and $T_{offsetDSC}$.

There are many factors involved in the LCST. More information could be obtained if the polymeric interactions between different chains were diminished, i.e., if the polymer concentration in water decreased. For that purpose, the concentration of the polymeric solutions was decreased to 1 wt.%. The idea was to observe if there were noticeable changes when the interactions between polymeric chains were reduced. Figure 5A shows the phase transitions of LCST for all the polymeric solutions (1 wt.%) whose shape is similar for all the samples. In this sense, hydrophobic transitions do not show differences probably because interactions between different polymeric chains are minimized.

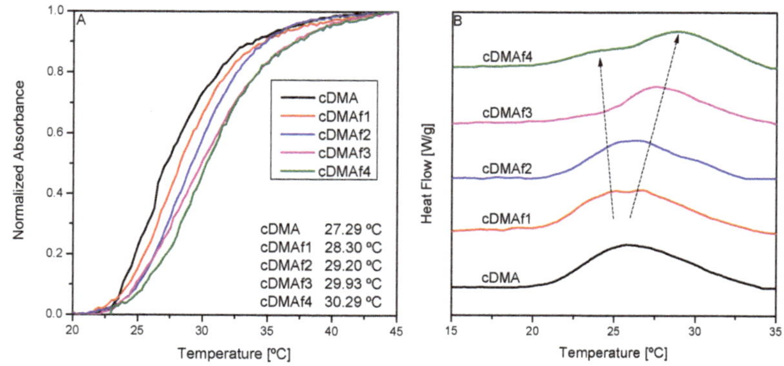

Figure 5. Phase transition temperatures analyzed by UV–vis spectroscopy (**A**) and calorimetry (**B**) for all polymeric solutions (1 wt.%).

Another important fact is associated with the trend drawn by the cloud points, which increase as organometallic moieties rise. The reproducibility of the experiments repeated three times only opens the possibility of stretching of the polymeric chain due to the hydrophobic forces of the organometallic complexes. However, for a mechanistic discussion, a much wider range of samples would be necessary.

Calorimetry could be a good tool for getting more information about these transitions as it is more sensitive than UV–vis spectroscopy. Figure 5B displays the calorimetric curves measured using a polymeric concentration in water of 1 wt.%. The pure copolymer exhibits a single transition which is subsequently divided into two parts as the content of the organometallic complexes rises. Both transitions are clearly distinguished for cDMAf3 and cDMAf4, showing that two phenomena are taking place. The effect could be associated with a stretching of the polymeric chains due to the presence of the organometallic compounds whose hydrophobic strength promoted by their length and volume over the main backbone could isolate the different polymeric sequences. Thus, this arrangement of the polymeric chains could induce a heterogeneous LCST-response regarding the pure copolymer where there are some interactions between the same polymeric chains (Scheme 3).

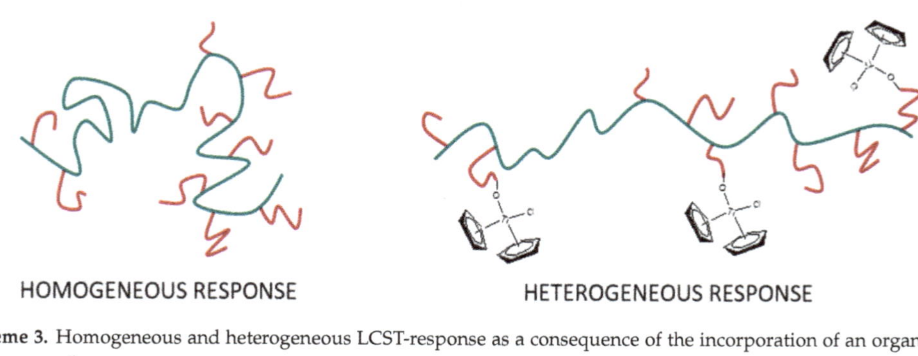

Scheme 3. Homogeneous and heterogeneous LCST-response as a consequence of the incorporation of an organometallic compound.

Transitions seem to be composed of the overlapping of two phenomena. The phenomenon or transition placed at lower temperature seems to decrease as the organometallic compound rises, while the highest one exhibits the opposite behavior, i.e., the peaks look displaced to higher temperatures. The effect at a lower temperature is clearly observed for the highest content of organometallic compound (cDMAf4), and could be related to the segments of the chains enriched in DMA and, consequently, in organometallic moieties. Then, it could be justified that the presence of these hydrophobic groups could move the transition to lower temperatures depending on the content of organometallics. Nevertheless, the second transition seems to follow the opposite trend, i.e., it increases as hydrophobic groups rise, probably due to the stretching of the polymeric chain as interactions with surrounding polymeric chains must be reduced.

Figure 6 shows experiments related to the wettability of the functionalized structures and also for the pure copolymer measured at different temperatures below and above of LCST. Specifically, the hydrophobicity on the surface was evaluated by the water contact angle, where important differences can be observed. First of all, a clear trend is defined by the content of the organometallic complex, which induces a higher hydrophobicity inducing higher angles. Similar behavior was observed for other samples where hydrophilic or/and hydrophobic groups were involved. For PNIPAM an increase in the hydrophobicity was reported, while a copolymer with hydrophilic monomer decreases the contact angle [44]. On the other hand, PNIPAM-modified styrene-butadiene rubber was also studied in terms of contact angle for different contents of NIPAM, where it was demonstrated how strongly the hydrophilic behavior could be influenced below the LCST [45].

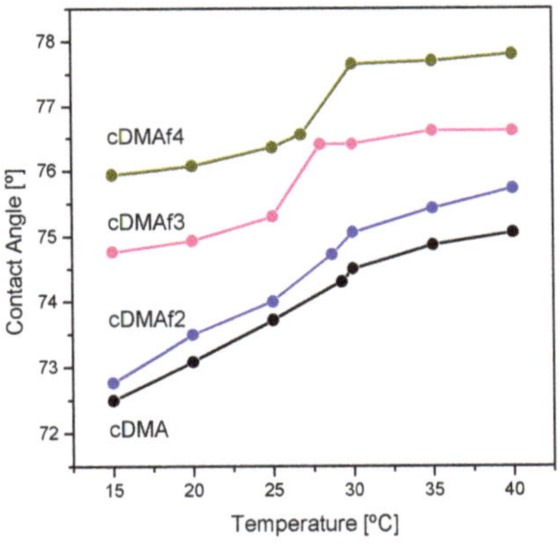

Figure 6. Contact angle values for functionalized samples and pure copolymer measured at different temperatures.

Specifically, cDMAf3 and cDMAf4 exhibit a clear jump around the LCST, which cannot be clearly detected for the pure copolymer or sample with a low amount of organometallic complex (cDMAf2). Nevertheless, these samples show a greater change between 15 °C and 40 °C as can be deduced from cDMAf2 where the angle increases around 3° while it is below 2° for cDMAf4. This fact could be associated with the presence of the organometallic complex, which could reduce the impact of the hydrophobic contributions of DMA below LCST as was observed for the hydrophobic transitions observed by calorimetry and UV–vis spectroscopy. The results fit well with the literature as remarkable changes are exclusively observed below LCST, while above the LCST, the angle remains constant [45].

The presence of the organometallic compound could play an important role along the transition LCST showing a clear jump which could be associated with the coil to globule transition where the organometallic moieties could be exposed outside of the globule due to its hydrophobic nature inducing those high angles. Nevertheless, this is a hypothesis, as the polymers are not dissolved in water.

Figure 7 exhibits the absorbance of the suspensions composed of different concentrations of DNA and functionalized materials (cDMA and cDMAf4). For both samples, the intensity of the peaks rises when DNA concentration increases. In addition, a hypsochromic effect was observed between the different DNA concentrations for each sample, as was reported previously [31].

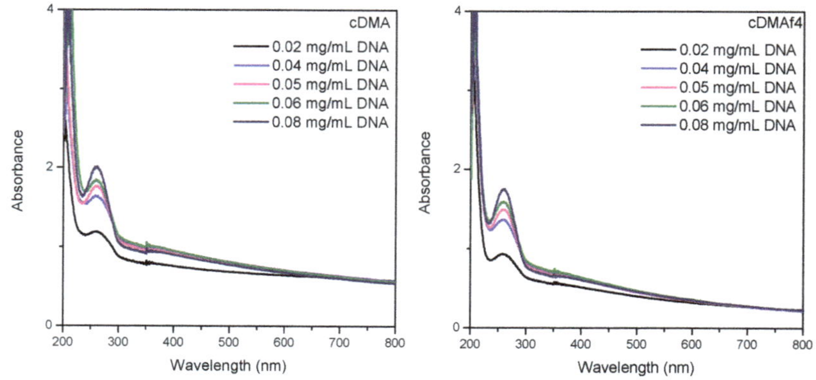

Figure 7. UV–vis spectra of DNA-interactions for cDMA (**left**) and cDMAf4 (**right**).

Both samples show the same trends, but it is important to indicate a change between the kind of peaks for the same DNA concentration but different samples. The peaks related to pure copolymer are slightly wider than for the functionalized sample cDMAf4, and consequently, this fact could be associated with the presence of the organometallic complex. Nevertheless, the experimental error could also be involved in these trends due to the low concentration of the organometallic complex.

4. Conclusions

The functionalization of copolymers based on N-isopropylacrylamide and dopamine methacrylamide allows high amounts of organometallic moieties incorporated in the polymeric chains to be obtained in comparison with other complex methods reported in the literature.

The functionalization of the polymers is carried out through the -OH groups, confirming the hypothesis reported in our previous publications about the functionalization routes. In this sense, the new copolymers allow getting information from FTIR spectroscopy which shows clear changes associated with ν(C–O) bonds in the band placed at 1253 cm^{-1}. The incorporation of an organometallic complex can affect the LCST behavior, but the polymer concentration in water will define those changes as interactions with surrounding polymeric chains will be modified; if there are polymeric interactions between different chains, those will be influenced by the presence and content of an organometallic complex. Those could affect the polymeric chains composed of pure NIPAM reducing its phase transition temperature associated with the LCST for these polymeric sequences and, consequently, obtaining a shorter hydrophobic LCST-transition. If the interactions between surrounding polymeric chains are minimized (low polymeric concentrations in water), the organometallic complex could reduce the interactions between parts of the same polymer chain. This fact could open a new perspective concerning controlling the LCST transition due to the presence and the content of organometallic complexes.

Contact angle shows the effect of the incorporation of an organometallic complex into the polymeric chains due to its promotion of a higher hydrophobicity. In addition, these contributions seem to modulate the change of temperature below and above LCST.

The DNA-interactions were tested for the resulting materials, showing that DNA can be affected probably due to electrostatic interactions.

Author Contributions: Conceptualization, A.G.-P.; methodology, M.M.-Z.; formal analysis, A.G.-P. and I.Q.-G.; investigation, M.M.-Z. and A.G.-P.; data curation, I.Q.-G. and A.G.-P.; writing—original draft preparation, A.G.-P.; writing—review and editing, F.J.S., I.Q.-G. and S.H.; supervision, A.G.-P. and V.S.-M.; funding acquisition, J.B., B.S. and I.Q.-G. All authors have read and agreed to the published version of the manuscript.

Funding: The authors want to thank the funding obtained from Ministerio de Ciencia, Innovación y Universidades (MCIU), Agencia Estatal de Investigación (AEI) and Fondo Europeo de Desarrollo Regional (FEDER, UE) through the project: PGC2018-095364-B-I00.

Institutional Review Board Statement: Not applicable.

Informed Consent Statement: Not applicable.

Data Availability Statement: Not applicable.

Conflicts of Interest: The authors declare no conflict of interest.

Abbreviations

LCST	Lower critical solution temperature
NIPAM	*N*-isopropylacrylamide
DMA	Dopamine methacrylamide
Cp_2TiCl_2	bis(cyclopentadienyl)titanium (IV) dichloride

References

1. Hirama, A.; Chou, L.-C.; Kakuchi, R. Streamlined access to end-functionalized thermoresponsive polymers via a combination of bulk RAFT polymerization and quasi solvent-free Passerini three-component reaction. *Polym. J.* **2021**, *53*, 1175–1185. [CrossRef]
2. Gosecki, M.; Ziemczonek, P.; Maczugowska, P.; Czaderna-Lekka, A.; Kozanecki, M.; Gosecka, M. The influence of 2-acrylamidephenylboronic acid on the phase behaviour of its copolymers with N-isopropylacrylamide in aqueous solution. *Polym. Chem.* **2021**, *12*, 3264–3275. [CrossRef]
3. Bratek-Skicki, A. Towards a new class of stimuli-responsive polymer-based materials—Recent advances and challenges. *Appl. Surf. Sci. Adv.* **2021**, *4*, 100068. [CrossRef]
4. Dinari, A.; Abdollahi, M.; Sadeghizadeh, M. Design and fabrication of dual responsive lignin-based nanogel via "grafting from" atom transfer radical polymerization for curcumin loading and release. *Sci. Rep.* **2021**, *11*, 1962. [CrossRef]
5. Howaili, F.; Ozliseli, E.; Kucukturkmen, B.; Razavi, S.M.; Sadeghizadeh, M.; Rosenholm, J.M. Stimuli-Responsive, Plasmonic Nanogel for Dual Delivery of Curcumin and Photothermal Therapy for Cancer Treatment. *Front. Chem.* **2020**, *8*, 602941. [CrossRef]
6. Zeweldi, H.G.; Bendoy, A.P.; Park, M.J.; Shon, H.K.; Kim, H.-S.; Johnson, E.M.; Kim, H.; Chung, W.-J.; Nisola, G.M. Supramolecular host-guest complex of methylated β-cyclodextrin with polymerized ionic liquid ([vbim]TFSI) as highly effective and energy-efficient thermo-regenerable draw solutes in forward osmosis. *Chem. Eng. J.* **2021**, *411*, 128520. [CrossRef]
7. Rahman, M.S.; Brown, J.; Murphy, R.; Carnes, S.; Carey, B.; Averick, S.; Konkolewicz, D.; Page, R.C. Polymer Modification of Lipases, Substrate Interactions, and Potential Inhibition. *Biomacromolecules* **2021**, *22*, 309–318. [CrossRef]
8. García-Peñas, A.; Wang, Y.; Mena-Palomo, I.; López-Collazo, E.; Díaz-García, D.; Gómez-Ruiz, S.; Stadler, F.J. Preparation, thermoresponsive behavior, and preliminary biological study of functionalized poly(N-isopropylacrylamide-co-dopamine methacrylamide) copolymers with an organotin(IV) compound. *Polym. Test.* **2021**, *94*, 107046. [CrossRef]
9. Xu, R.; Wu, Y.; Ma, Y.; Zhang, Y.; Ma, S.; Cai, M.; Zhou, F.; Liu, W. Synthesis of novel copolymer/Mn+ biomimetic adhesives based on temperature strengthened coacervation effect. *Chem. Eng. J.* **2021**, *425*, 127249. [CrossRef]
10. Liang, W.; García-Peñas, A.; Sharma, G.; Kumar, A.; Stadler, F.J. Competition between Physical Cross-Linking and Phase Transition Temperature in Blends Based on Poly(N-isopropylacrylamide-co-N-ethylacrylamide) Copolymers and Carboxymethyl Cellulose. *Macromol. Chem. Phys.* **2020**, *221*, 2000081. [CrossRef]
11. Jonnalagadda, U.S.; Nguyen, T.M.; Li, F.; Lee, J.H.C.; Liu, X.; Goto, A.; Kwan, J.J. Sol–Gel Transitions of Comb-Like Polymethacrylate Copolymers by Mechano-Thermal Stimuli in Water. *Macromol. Chem. Phys.* **2020**, *221*, 202000088. [CrossRef]
12. Du, B.; Stadler, F.J. Functional Polymer Solutions and Gels-Physics and Novel Applications. *Polymers* **2020**, *12*, 676. [CrossRef] [PubMed]
13. Khodeir, M.; Antoun, S.; Ruymbeke, E.; Gohy, J.F. Temperature and Redox-Responsive Hydrogels Based on Nitroxide Radicals and Oligoethyleneglycol Methacrylate. *Macromol. Chem. Phys.* **2020**, *221*, 1900550. [CrossRef]
14. Bongiovanni Abel, S.; Riberi, K.; Rivarola, C.R.; Molina, M.; Barbero, C.A. Synthesis of a Smart Conductive Block Copolymer Responsive to Heat and Near Infrared Light. *Polymers* **2019**, *11*, 1744. [CrossRef] [PubMed]
15. GhavamiNejad, A.; Hashmi, S.; Vatankhah-Varnoosfaderani, M.; Stadler, F.J. Effect of H2O and reduced graphene oxide on the structure and rheology of self-healing, stimuli responsive catecholic gels. *Rheol. Acta* **2016**, *55*, 163–176. [CrossRef]

16. Abd-El-Aziz, A.S.; Manners, I. (Eds.) Frontiers in Transition Metal-Containing Polymers. *J. Inorg. Organomet. Polym. Mater.* **2007**, *17*, 687–688. [CrossRef]
17. Whittell, G.R.; Manners, I. Metallopolymers: New Multifunctional Materials. *Adv. Mater.* **2007**, *19*, 3439–3468. [CrossRef]
18. Bellas, V.; Rehahn, M. Polyferrocenylsilane-based polymer systems. *Angew. Chem. Int. Ed. Engl.* **2007**, *46*, 5082–5104. [CrossRef]
19. Branham, K.E.; Mays, J.W.; Gray, G.M.; Sanner, R.D.; Overturf, G.E.; Cook, R. Reactions of Titanocene Derivatives with Molecular Carboxylic Acids and Copolymers Bearing Carboxylic Acid Groups. *Appl. Organomet. Chem.* **1997**, *11*, 213–221. [CrossRef]
20. Tomita, I.; Ueda, M. Synthesis and Reactions of Organometallic Polymers Containing Titanacyclobutene Units in the Main Chain. *J. Inorg. Organomet. Polym. Mater.* **2006**, *15*, 511–518. [CrossRef]
21. Pentela, N.; Murugan, P.; Jaisankar, S.N.; Samanta, D.; Mandal, A.B. Immobilization of ruthenium benzylidene on thermoresponsive polymer: Methodology and application. *J. Organomet. Chem.* **2015**, *778*, 42–46. [CrossRef]
22. Pentela, N.; Gayathri, V.; Samanta, D. N-heterocyclic carbene bearing thermoresponsive poly(NIPAM) supported palladium (II) complex: Synthetic strategy and application. *J. Organomet. Chem.* **2020**, *913*, 121196. [CrossRef]
23. Du, W.; Slany, M.; Wang, X.; Chen, G.; Zhang, J. The Inhibition Property and Mechanism of a Novel Low Molecular Weight Zwitterionic Copolymer for Improving Wellbore Stability. *Polymers* **2020**, *12*, 708. [CrossRef] [PubMed]
24. Clavel, C.M.; Păunescu, E.; Nowak-Sliwinska, P.; Dyson, P.J. Thermoresponsive organometallic arene ruthenium complexes for tumour targeting. *Chem. Sci.* **2014**, *5*, 1097–1101. [CrossRef]
25. Lanzalaco, S.; Armelin, E. Poly(N-isopropylacrylamide) and Copolymers: A Review on Recent Progresses in Biomedical Applications. *Gels* **2017**, *3*, 36. [CrossRef] [PubMed]
26. Vatankhah-Varnoosfaderani, M.; GhavamiNejad, A.; Hashmi, S.; Stadler, F.J. Hydrogen bonding in aprotic solvents, a new strategy for gelation of bioinspired catecholic copolymers with N-isopropylamide. *Macromol. Rapid Commun.* **2015**, *36*, 447–452. [CrossRef]
27. GhavamiNejad, A.; Sasikala, A.R.K.; Unnithan, A.R.; Thomas, R.G.; Jeong, Y.Y.; Vatankhah-Varnoosfaderani, M.; Stadler, F.J.; Park, C.H.; Kim, C.S. Mussel-Inspired Electrospun Smart Magnetic Nanofibers for Hyperthermic Chemotherapy. *Adv. Funct. Mater.* **2015**, *25*, 2867–2875. [CrossRef]
28. Amirian, J.; Zeng, Y.; Shekh, M.I.; Sharma, G.; Stadler, F.J.; Song, J.; Du, B.; Zhu, Y. In-situ crosslinked hydrogel based on amidated pectin/oxidized chitosan as potential wound dressing for skin repairing. *Carbohydr. Polym.* **2021**, *251*, 117005. [CrossRef]
29. Zhu, M.Q.; Wang, L.Q.; Exarhos, G.J.; Li, A.D. Thermosensitive gold nanoparticles. *J. Am. Chem. Soc.* **2004**, *126*, 2656–2657. [CrossRef]
30. Du, L.; Biswas, C.S.; Wu, Y.; GhavamiNejad, A.; Stadler, F.J. Small and large amplitude oscillatory shear behavior of supramolecular gels based on dopamine-boronic acid interactions. *J. Rheol.* **2019**, *63*, 391–404. [CrossRef]
31. Wang, Y.; García-Peñas, A.; Gómez-Ruiz, S.; Stadler, F.J. Surrounding Interactions on Phase Transition Temperature Promoted by Organometallic Complexes in Functionalized Poly(N-isopropylacrylamide-co-dopamine methacrylamide) Copolymers. *Macromol. Chem. Phys.* **2020**, *221*, 2000035. [CrossRef]
32. Glass, P.; Chung, H.; Washburn, N.R.; Sitti, M. Enhanced reversible adhesion of dopamine methacrylamide-coated elastomer microfibrillar structures under wet conditions. *Langmuir* **2009**, *25*, 6607–6612. [CrossRef] [PubMed]
33. Garcia-Penas, A.; Biswas, C.S.; Liang, W.; Wang, Y.; Yang, P.; Stadler, F.J. Effect of Hydrophobic Interactions on Lower Critical Solution Temperature for Poly(N-isopropylacrylamide-co-dopamine Methacrylamide) Copolymers. *Polymers* **2019**, *11*, 991. [CrossRef] [PubMed]
34. García-Peñas, A.; Biswas, C.S.; Liang, W.; Wang, Y.; Stadler, F.J. Lower Critical Solution Temperature in Poly(N-Isopropylacrylamide): Comparison of Detection Methods and Molar Mass Distribution Influence. *Macromol. Chem. Phys.* **2019**, *220*, 1900129. [CrossRef]
35. Vatankhah-Varnoosfaderani, M.; Hashmi, S.; GhavamiNejad, A.; Stadler, F.J. Rapid self-healing and triple stimuli responsiveness of a supramolecular polymer gel based on boron–catechol interactions in a novel water-soluble mussel-inspired copolymer. *Polym. Chem.* **2014**, *5*, 512–523. [CrossRef]
36. GhavamiNejad, A.; Hashmi, S.; Joh, H.I.; Lee, S.; Lee, Y.S.; Vatankhah-Varnoosfaderani, M.; Stadler, F.J. Network formation in graphene oxide composites with surface grafted PNIPAM chains in aqueous solution characterized by rheological experiments. *Phys. Chem. Chem. Phys.* **2014**, *16*, 8675–8685. [CrossRef]
37. Hashmi, S.; GhavamiNejad, A.; Stadler, F.J.; Wu, D. Oscillations in modulus in solutions of graphene oxide and reduced graphene oxide with grafted poly-N-isopropylamide. *Soft Matter.* **2015**, *11*, 1315–1325. [CrossRef]
38. Satoh, K.; Ito, D.; Kamigaito, M. Periodic Introduction of Water-Tolerant Titanatrane Complex to Poly(NIPAM) Prepared by Simultaneous Step-Growth and Living Radical Polymerization. In *Controlled Radical Polymerization: Materials*; American Chemical Society: Washington, DC, USA, 2015; Volume 1188, pp. 1–14.
39. Haraguchi, K.; Xu, Y. Thermal analyses of poly(N-isopropylacrylamide) in aqueous solutions and in nanocomposite gels. *Colloid Polym. Sci.* **2012**, *290*, 1627–1636. [CrossRef]
40. Chung, J.E.; Yokoyama, M.; Suzuki, K.; Aoyagi, T.; Sakurai, Y.; Okano, T. Reversibly thermo-responsive alkyl-terminated poly(N-isopropylacrylamide) core-shell micellar structures. *Colloids Surf. B Biointerfaces* **1997**, *9*, 37–48. [CrossRef]
41. Patterson, D. Free Volume and Polymer Solubility. A Qualitative View. *Macromolecules* **2002**, *2*, 672–677. [CrossRef]
42. Otake, K.; Inomata, H.; Konno, M.; Saito, S. Thermal analysis of the volume phase transition with N-isopropylacrylamide gels. *Macromolecules* **2002**, *23*, 283–289. [CrossRef]

43. Song, J.M.; Winnik, F.M.; Brash, J.L. Synthesis and Solution Properties of Fluorescently Labeled Amphiphilic (N-alkylacrylamide) Oligomers. *Macromolecules* **1998**, *31*, 109–115. [CrossRef]
44. Sanz, B.; von Bilderling, C.; Tuninetti, J.S.; Pietrasanta, L.; Mijangos, C.; Longo, G.S.; Azzaroni, O.; Giussi, J.M. Thermally-induced softening of PNIPAm-based nanopillar arrays. *Soft Matter.* **2017**, *13*, 2453–2464. [CrossRef] [PubMed]
45. Hermann, A.; Mruk, R.; Roskamp, R.F.; Scherer, M.; Ma, L.; Zentel, R. Poly(N-isopropylacrylamide)-Modified Styrene-Butadiene Rubber as Thermoresponsive Material. *Macromol. Chem. Phys.* **2014**, *215*, 32–43. [CrossRef]

polymers

MDPI

Article

Assessment of Toxicity and Biodegradability of Poly(vinyl alcohol)-Based Materials in Marine Water

Olalla Alonso-López [1], Sara López-Ibáñez [1] and Ricardo Beiras [1,2,*]

1 Grupo ECOTOX (EcoCost) ECIMAT, Centro de Investigación Mariña, Universidade de Vigo, 36331 Galicia, Spain; olallacristina.alonso@uvigo.es (O.A.-L.); salopez@uvigo.es (S.L.-I.)
2 Departamento de Ecoloxía e Bioloxía Animal, Facultade de Bioloxía, Universidade de Vigo, 36310 Galicia, Spain
* Correspondence: rbeiras@uvigo.es

Abstract: Due to the continuous rise in conventional plastic production and the deficient management of plastic waste, industry is developing alternative plastic products made of biodegradable or biobased polymers. The challenge nowadays is to create a new product that combines the advantages of conventional plastics with environmentally friendly properties. This study focuses on the assessment of the potential impact that polyvinyl alcohol (PVA)-based polymers may have once they are released into the marine environment, in terms of biodegradation in seawater (assessed by the percentage of the Theoretical Oxygen Demand, or % ThOD, of each compound) and aquatic toxicity, according to the standard toxicity test using *Paracentrotus lividus* larvae. We have tested three different materials: two glycerol-containing PVA based ones, and another made from pure PVA. Biodegradation of PVA under marine conditions without an acclimated inoculum seems to be negligible, and it slightly improves when the polymer is combined with glycerol, with a 5.3 and 8.4% ThOD achieved after a period of 28 days. Toxicity of pure PVA was also negligible (<1 toxic units, TU), but slightly increases when the material included glycerol (2.2 and 2.3 TU). These results may contribute to a better assessment of the behavior of PVA-based polymers in marine environments. Given the low biodegradation rates obtained for the tested compounds, PVA polymers still require further study in order to develop materials that are truly degradable in real marine scenarios.

Keywords: poly(vinyl alcohol); glycerol; microplastics; biodegradation; toxicity; marine water

Citation: Alonso-López, O.; López-Ibáñez, S.; Beiras, R. Assessment of Toxicity and Biodegradability of Poly(vinyl alcohol)-Based Materials in Marine Water. *Polymers* **2021**, *13*, 3742. https://doi.org/10.3390/polym13213742

Academic Editors: Florian J. Stadler and Alberto García-Peñas

Received: 30 September 2021
Accepted: 26 October 2021
Published: 29 October 2021

1. Introduction

1.1. Marine Plastic Pollution

Plastics are one of the most used materials nowadays due to their low cost, light weight, high durability and excellent isolation properties. The annual production in 2019 reached almost 370 million tons [1]. Nevertheless, this high demand and long environmental persistence caused their widespread accumulation into the environment. Plastic is the predominant component of marine litter; it is estimated that around 8 million tons of mismanaged plastic are released into the ocean every year [2]. Environmental weathering of plastics leads to fragmentation into smaller particles, and fractions less than 5 mm are called microplastics [3]. Microplastics pose a potential risk not only for filter feeders, which can end up consuming plastic debris [4] but also for all trophic levels [5]. As particle size decreases, the bioavailability, translocation and toxicity to organisms increases [6].

United Nations Sustainable Development Goals 2015 (SDGs) include, as part of Goal 14, a call to reduce impacts from plastics [7]. Efforts from all around the world are now developing new synthetic materials innocuous for the environment according to the different end-of-life scenarios. This includes regulatory efforts such as the EU initiative to phase out some single-use plastics (Directive 2019/904). Materials based on renewable components and biodegradable options are already being developed as alternatives to conventional polymers. Nowadays, the production of plastics from renewable raw materials is 1% of total plastic production [8]. However, although these new materials are promising, they present new challenges, such as the potential toxicity of the degradation products, or their rapid fragmentation into microplastics [9].

1.2. Polyvinyl Alcohol

Poly(vinyl alcohol) (PVA) is a synthetic water-soluble biopolymer that is generally prepared by the saponification of polyvinyl acetate [10]. Remarkable advantages of this potentially biodegradable material are transparency, good mechanical and thermal properties and resistance to oxygen permeation [11].

Products made by PVA constitute the largest volume of water-soluble polymer produced this century [12], with 650,000 tons of PVA per year around the globe [13]. PVA can be found in a wide range of items, including food packaging (31.4% of the demand), construction, electronics, coatings, printing, textile, cosmetics and paper [14]. Likewise, the use of PVA in pharmaceutical and medical applications, including tablets, contact lenses and surgical threads, have been reported [15]. Most of these materials can reach the environment without passing through any integrated system of waste treatment [12]. Another less known use of PVA that implies direct input into the ocean is the equipment for sports fishing, such as bags with dry fishing bait attached to the hook that are eventually released [16].

PVA is perhaps the only fully synthetic C-chain polymer biodegradable in aerobic [17] and anaerobic conditions [18], although in laboratory trials, acclimation in order to allow growth of selected microorganisms is usually essential to achieve rapid degradation [17]. Certain bacteria and fungi contain specific oxidases and dehydrogenases that promote an oxidative reaction of the tertiary carbon atoms of PVA chain forming hydrolyzable b-hydroxylketone and 1,3-diketone groups along the polymer backbone [19,20]. Nevertheless, these microorganisms are not present in most environmental scenarios where degradation rates are very low [21].

1.3. Plastic Biodegradation

Biodegradability is a plastic end-of-life option that uses the microorganisms present in a particular environment to completely remove plastic products by mineralization to CO_2, water and biomass. This process depends on the chemical structure of the polymer, the additives that it contains and the environmental conditions. Usually, the biodegradation of plastics is carried out by scission of the polymer backbone by hydrolysis or enzymatic cleavage initiated by microorganisms that can digest the polymer [22]. Biodegradable plastics can be decomposed on industrial composting facilities [23,24] but many of them are not suitable for biodegradation in natural environments. Previous studies tested the biodegradability of potentially biodegradable plastics in seawater by recording weight loss [25–27], oxygen consumption as biological oxygen demand (BOD) or CO_2 evolution [28,29].

1.4. Toxicity of Plastics

Marine pollution analysis was commonly based on the performance of chemical analysis. Nonetheless, due to the high complexity of pollutant mixtures and their interactions with environmental factors, this does not offer sufficient information on their potential effects on organisms in the natural environment. Therefore, the evaluation of toxicity in marine waters must integrate conventional analyses with biological methods such as bioassays [30].

The toxicity of common plastic in the marine environment was previously studied [31–33], and frequently associated to chemical additives [34] or sorbed environmental pollutants [33]. Nonetheless, biopolymer toxicity is an emerging issue, since adverse effects of biodegradation products must be taken into account. Some components of water-soluble biopolymers such as ethylene released during degradation can have negative effects on surrounding organisms [35]. On the other hand, some biopolymers, e.g., PHB, seemed to impact aquatic organisms through different mechanisms associated with the higher abundance of plastic particles within the nanometric range found in these resins and absent in other materials [34,36].

The main objective of this research is to provide new insights about the potential risks of microplastics from PVA materials for the marine environment. With that aim, biodegradation studies and ecotoxicological tests were performed in order to have a complete insight into the potential impact of these materials.

2. Materials and Methods

2.1. Tested Materials

PVA samples were provided by the Plastic Technology Centre AIMPLAS (Spain) from three different stocks (Table 1). Two of them consisted of a mixture of PVA and glycerol, a common plasticizer used in the production of bioplastic [37], and the third one was a plain PVA resin. All the materials were micronized by a ZM200 ultracentrifuge mill (Retsch, Verder Scientific) and sieved through a <250 μm metallic mesh in order to standardize particle size for further testing. A biodegradable material, PHB powder (<250 μm), purchased from Helian Polymers, was used as a reference material.

Table 1. Characteristics of the materials tested.

Reference	Brand	Hydrolysis, Mole %	Viscosity (cps)	% PVA	% Glycerol
PVA.029	Selvol™ Polyvinyl Alcohol 205	87.00–89.00	5.2–6.2	85	15
PVA.030	Exceval HR 3010	99.0–99.4	12.0–16.0	85	15
PVA.031	PVA KURARAY POVALTM	98.0–99.0	3.2–3.8	100	0

2.2. Biodegradation Study

Biodegradation of the materials was tested following the UNE-EN ISO 14851:2019 [38] for the oxygen demand measurement in a close respirometer, adapted to seawater as follows. $NaNO_3$, $Na_2HPO_4{\cdot}2H_2O$, and $FeCl_3{\cdot}6H_2O$ purchased from Merck were used as sources of N, P and Fe, and the nutrient formulation was calculated to fit the well-known Redfield ratio 106:16:1:0.1 for C:N:P:Fe [39,40].

The tests were carried out with a set of 500 mL amber glass incubation bottles closed by respirometer units with piezoresistive electronic pressure sensors OxiTop®-i IS 6-WTW. The experiments included blanks (with no sample), PHB positive controls and the tested plastic materials, all treatments per duplicate. The medium for the experiment included 0.8 μm filtered seawater sterilized with UV light, enriched with nutrients as explained above, and 1% v. of marine inoculum. The inoculum used was sediment interstitial water collected the same testing day from the beach of Canido (Vigo, NW Iberian Peninsula) in a location under the influence of a freshwater course submitted to frequent discharges of urban wastewater. Then, 250 mL of the medium and 25 mg of microplastic sample were added to each bottle with a magnetic stirrer for continuous mixing. Before closing the bottles, a rubber with KOH pearls was placed in the neck in order to capture the evolved CO_2. Biological oxygen demand (expressed as mg O_2/L) was recorded every day for a 28-day period.

Blank-corrected BOD values are expressed as % of the Theoretical Oxygen Demand (ThOD), defined as the theoretical amount of oxygen required to fully degrade the whole organic carbon content of the polymer to CO_2, as described in UNE-EN ISO 14851:2019 [38], and as percentage of the BOD recorded in the positive control (%C+). The latter is more suited to biopolymers, since the main purpose is to detect relevant biodegradation compared to a well-known biodegradable polymer, and because the actual atomic composition of tested polymers is not always disclosed.

2.3. Toxicity Test

Marine aquatic toxicity of PVA leachates was tested using the *Paracentrotus lividus* sea-urchin embryo test (SET) according to [41]. The leachates were obtained in artificial seawater [42] at 10 g/L and tested using geometrical serial dilutions (×1, ×1/3, ×1/10. . .) until the dilution was found, with no toxic effect (NOEC). Four replicates per treatment and eight controls were added. Sea urchins were provided by the Marine Culture Unit of ECIMAT (CIM-University of Vigo) the same day the tests were performed. Eggs were fertilized with a few µL of sperm in a 50 mL measuring cylinder and transferred before the first division into glass vials with 4 mL of the experimental solution and a density of 40 individuals per mL. Vials were closed with Teflon-lined caps and kept at 20 °C in the dark for 48 h. Afterwards, the incubation time vials were fixed with 40% formalin for ulterior observation under an inverted microscope (Leica DMI 4000B). The length as a maximum linear dimension was measured on 35 larvae per vial using Leica Application Suite LAS image analysis software version 4.12.0 (Leica Microsystems, Germany). The acceptability criteria for this test were the percentage of fertilized eggs (>98%) and size increase in controls >253 µm [43].

Statistical analyses were carried out using IBM SPSS (v. 24). Normal distribution and homoscedasticity were checked using the Shapiro–Wilk and Levene's tests, respectively. Leachate dilutions that produced larval growths significantly different to the control ($p < 0.05$) were identified using Dunnett's post hoc test or, when the variances were not homogeneous, Dunnett's T3 test in order to find the lowest no observed adverse effects concentration (NOEC) and the lowest observed adverse effect concentration (LOEC). The leachate dilutions that produced a 50% decrease in larval growth with respect to the control (EC_{50}) was also calculated and their 95% confidence intervals (CIs) were calculated adjusting the data obtained to the Probit dose-response model. Toxic Units (TU) were calculated as TU= $1/EC_{50}$, and materials were classified following the assessment criteria shown in Table 2.

Table 2. Toxicity classification based on the sea-urchin embryo test Toxic Units (TU).

EC_{50} (mg/L)	TU	Toxicity
>10,000	<1	None
2000–10,000	$1 \leq$ TU< 5	Slight
400–2000	$5 \leq$ TU< 25	Relevant
<400	$\geq$25	High

3. Results and Discussion

3.1. Biodegradability Test

Figure 1 shows the biodegradation results for the 28 days of incubation obtained for each material. According to UNE-EN ISO 14851:2019 guidelines [38], a substance is considered biodegradable if the BOD is higher than 60% ThOD. PHB was used as positive control due to its well-known biodegradability in seawater, as shown by Tachibana et al. who reported 80% biodegradation in 25 days [28]. Similar percentages of biodegradation were found in the present study, where the PHB control achieved 70% degradation after 28 days. Some standards propose an extended incubation period of 60 days, and up to 180 days (ISO 19679, ISO 23977-2, ASTM D7991-15) [44,45]. This extension considerably increases the technical complexity of the experimental setup and increases the chances of failures during the execution of the tests. Moreover, even though we may need to consider a longer period for certain materials, it was observed that the degradation became stable after 28 days for both positive control and tested materials. Materials PVA.029 and PVA.030, including 15% glycerol in their composition, showed a final biodegradation percentage of 5.3 and 8.4% ThOD, respectively, and are thus classified as slightly biodegradable. For the PVA.031 sample, composed by PVA only, we observed that the biodegradation was negligible. However, studies about degradability of PVA on freshwater inoculated with municipal sewage sludge reported percentages of biodegradation of 13% during 21 days [12]. The marine medium used in the present studies (pH 8.3) may lack microbial strains present in wastewater. Additionally, seawater shows a pH > 8, which may have retarded degradation, since biodegradation of PVA was reported to be higher in acidic aqueous solutions than in alkaline ones [46].

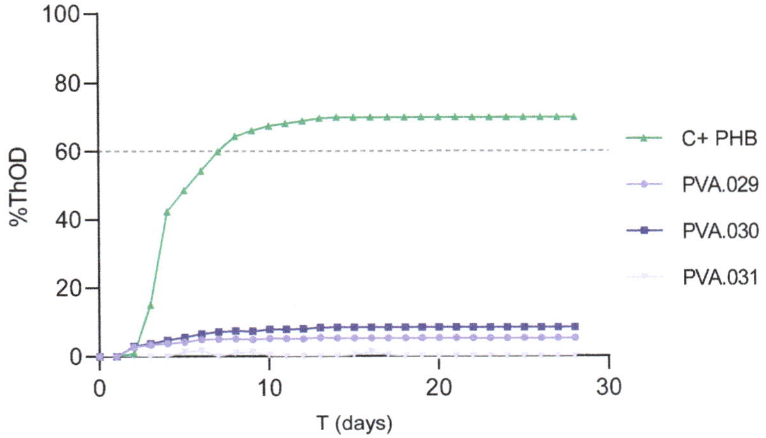

Figure 1. Biodegradation expressed by %ThOD for the different materials and the positive control (C+ PHB). Discontinuous line: 60% biodegradability threshold.

Expressed as a percentage of the positive control, the biodegradation rate of the sample PVA.029 was 7.8 and 12.4% for the 030.PVA one, values that still represent a low degree of biodegradation. As expected, the material with the higher degree of hydrolysis (030.PVA) results as the most biodegradable, as reduced molecular weight is a precondition for microbial attack, and the hydrolytic mechanisms enhance the biodegradation processes [47]. Moreover, PVA materials can reach up to 60% degradation in 32 days depending on the degree of solubility of the polymer [48].

The detected increase of 6.85% on average in the biodegradation rate for glycerol-containing PVA may be due to the changes in the hydrophilic characteristics of the glycerol that reduce internal hydrogen bonds in the polymer chain and decrease the residual mass as described by Abdullah and Dong (2019) [49]. They observed an increase of 23.33% in biodegradation rates when adding glycerol. Moreover, raw glycerol has biodegradation on natural water between 68 and 78% [50].

It should be noted that due to the lack of research about biodegradability on microparticles from bioplastics, this study can be only compared to studies using larger fragments (films, pellets, etc.). There is a gap of information in this area that needs to be covered, given that the final faith of microplastics and their degradation products is not well known [51].

Some labels have been developed by industry to distinguish plastics that can biodegrade in the environment. For instance, the label created by Vinçotte OK Biodegradable WATER applies the standard BS EN ISO 14851:2019 [38,52]. Nevertheless, as Harrison et al. pointed out, there is no agreement on which standards to use for plastic biodegradation [53], and a more realistic point of view unifying all the criteria required for these specific materials is needed. It is also important to bear in mind that biodegradability is an intrinsic property of a material, but performed biodegradation will necessarily depend on environmental conditions, and thus no single standard will be able to be representative of the multiple end-of-life scenarios present in the aquatic environments.

3.2. Toxicity Test

Toxicity of the materials using the standard sea urchin model with a concentration of 10 g/L is reflected in Table 3. According to the current EU classification of chemicals, based on their short-term aquatic toxicity [54], all samples are classified as harmless (EC_{50} > 100 mg/L). Following the TU classification (see Table 2), the polymer composed of plain PVA showed a total absence of toxicity with a TU < 1. This result is supported by toxicity bioassays in seven other aquatic species including marine water individuals *Hyalella azteca* and *Leptocheirus plumulosus* that showed a low toxicity for the PVA material [55]. The materials PVA.029 and PVA.030 presented a certain degree of aquatic toxicity (2.3 and 2.2 TU, respectively) with a larval growth lower than the PVA.031 sample (Figure 2). Consequently, PVA.029 and PVA.030 could be classified as slightly toxic.

Table 3. Toxicity of different materials expressed as no observed adverse effects concentration (NOEC) and the lowest observed adverse effect concentration (LOEC); toxic units (TU); 50% decrease in larval growth with respect to the control (EC_{50}).

Sample	NOEC (g/L)	LOEC (g/L)	TU	CE_{50} (mg/L)
PVA.029	1	3.33	2.3	4285 (1585–12164)
PVA.030	1	3.33	2.2	4382 (3750–5017)
PVA.031	3.33	10	<1	2403.8 (1814.8–3773.5)

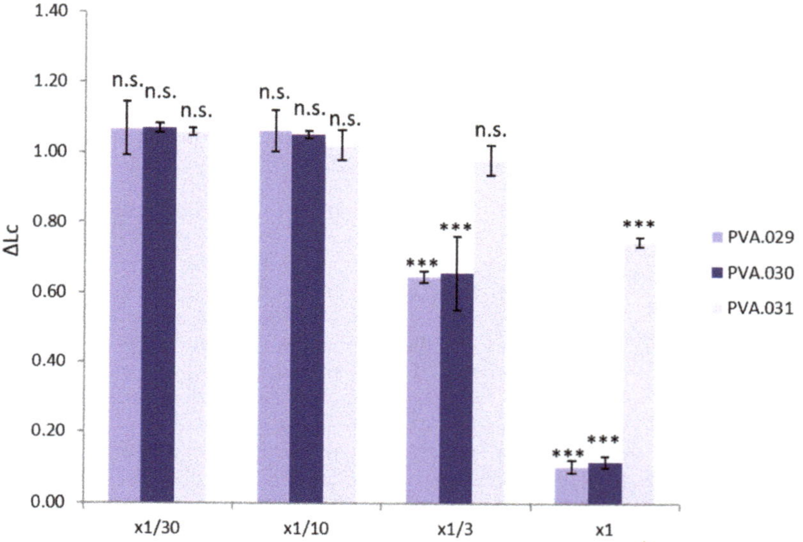

Figure 2. *Paracentrotus lividus* larvae size increase ratio compared to control treatment (ΔLc) for the different sample dilutions. * $p < 0.05$, ** $p < 0.01$, *** $p < 0.001$.

ECHA [56] shows a CE_{50} for glycerol on freshwater invertebrate organisms at 48 h of 1995 (1851 to 2068) mg/L, corresponding to a TU < 1. According to studies carried out on three different aquatic bioindicators, glycerol ethers can be classified as harmless for the environment in acute exposure [57]. Glycerol has been presented as a sustainable solvent for its good biodegradability and low toxicity on certain organisms [58], but its effects when blended with polymers and other materials need further research in order to confirm that these compounded materials do not pose a threat for the environment. In fact, we have found a slight but significant increase in toxicity for glycerol-PVA blends, compared to plain PVA.

Similarly, it is necessary to include the assessment of the potential toxicity of degradation products in the scheme of evaluation of biopolymers before they can be presented as more environmentally friendly alternatives compared to conventional plastics.

4. Conclusions

This study characterized the biodegradation and ecotoxicity of PVA polymers in marine environments. The results obtained, though limited, support the lack of biodegradability of PVA materials under conditions representative of a natural marine environment. The slight biodegradability of the blended materials was attributable to the glycerol component of the blend. Since plastics of different nature are increasingly found in marine environments, standards with a view to assess biodegradation under realistic marine end-of-life scenarios are urgently needed.

On the other hand, none of the tested polymers pose a relevant risk to the model marine organism used, the sensitive Sea urchin Embryo Test (SET), but slight toxicity arises for glycerol-containing blends. These tests should be conducted with a broader group of aquatic species. These standards should take into account not only biodegradability in terms of mineralization to CO_2, but also mechanical degradation, potential release of microplastics and lack of toxicity of additives and biodegradation products.

Author Contributions: Conceptualization, O.A.-L., S.L.-I. and R.B.; data curation, S.L.-I.; formal analysis, O.A.-L.; funding acquisition, R.B.; investigation, O.A.-L., S.L.-I. and R.B.; methodology, O.A.-L., S.L.-I. and R.B.; project administration, R.B.; resources, O.A.-L., S.L.-I. and R.B.; supervision, R.B.; writing—original draft, O.A.-L.; writing—review and editing, O.A.-L., S.L.-I. and R.B. All authors have read and agreed to the published version of the manuscript.

Funding: This research was supported by the RisBioPlas Project, financed by the Ministry of Science and Innovation—Spanish Agency of Research (AEI: 10.13039/501100011033); by LABPLAS Project (LAnd-Based Solutions for PLAStics in the Sea, Grant Agreement Nr.: 101003954, financed by the EU H2020 program); by the RESPONSE Project (toward a risk-based assessment of microplastic pollution in marine ecosystems, financed by the Joint Programming Initiative, Healthy and Productive Seas and Oceans, JPI Oceans-Spanish National Research funding Agency: PCI2020-112110). S.L.I. was supported by the Ministry of Universities of Spain (FPU grant reference FPU19/02280).

Institutional Review Board Statement: Not applicable.

Informed Consent Statement: Not applicable.

Data Availability Statement: Not applicable.

Acknowledgments: The authors thank the Plastic Technology Centre AIMPLAS (Valencia, Spain) for supplying the different PVA materials. Funding for open access charge: Universidade de Vigo/CISUG.

Conflicts of Interest: The authors declare no conflict of interest.

References

1. PlasticsEurope. Plastics—The Facts 2020. An Analysis in European Plastic Production, Demand and Waste Data. *PlasticEurope* **2020**, 1–64.
2. Napper, I.E.; Thompson, R.C. Plastic Debris in the Marine Environment: History and Future Challenges. *Glob. Chall.* **2020**, *1900081*, 1–9. [CrossRef]
3. Andrady, A.L. Microplastics in the Marine Environment. *Mar. Pollut. Bull.* **2011**, *62*, 1596–1605. [CrossRef]
4. Tanaka, K.; Takada, H. Microplastic Fragments and Microbeads in Digestive Tracts of Planktivorous Fish from Urban Coastal Waters. *Sci. Rep.* **2016**, *6*, 1–8. [CrossRef]
5. Barboza, L.G.A.; Dick Vethaak, A.; Lavorante, B.R.B.O.; Lundebye, A.K.; Guilhermino, L. Marine Microplastic Debris: An Emerging Issue for Food Security, Food Safety and Human Health. *Mar. Pollut. Bull.* **2018**, *133*, 336–348. [CrossRef]
6. Beiras, R.; Schönemann, A.M. Currently Monitored Microplastics Pose Negligible Ecological Risk to the Global Ocean. *Sci. Rep.* **2020**, *10*, 1–9. [CrossRef]
7. Walker, T.R. (Micro) Plastics and the UN Sustainable Development Goals. *Curr. Opin. Green Sustain. Chem.* **2021**, *30*, 100497. [CrossRef]
8. European Bioplastics. Bioplastics Market Data 2019. Available online: https://www.european-bioplastics.org/market/ (accessed on 1 September 2021).
9. Hidalgo-Ruz, V.; Gutow, L.; Thompson, R.C.; Thiel, M. Microplastics in the Marine Environment: A Review of the Methods Used for Identification and Quantification. *Environ. Sci. Technol.* **2012**, *46*, 3060–3075. [CrossRef]
10. Finch, C.A. Polyvinyl Alcohol, Properties and Applications. *J. Polym. Sci. Polym. Lett. Ed.* **1974**, *12*, 105–106. [CrossRef]
11. Kanatt, S.R.; Makwana, S.H. Development of Active, Water-Resistant Carboxymethyl Cellulose-Poly Vinyl Alcohol-Aloe Vera Packaging Film. *Carbohydr. Polym.* **2020**, *227*, 115303. [CrossRef]
12. Chiellini, E.; Corti, A.; Solaro, R. Biodegradation of Poly(Vinyl Alcohol) Based Blown Films under Different Environmental Conditions. *Polym. Degrad. Stab.* **1999**, *64*, 305–312. [CrossRef]

13. Xu, S.; Malik, M.A.; Qi, Z.; Huang, B.T.; Li, Q.; Sarkar, M. Influence of the PVA Fibers and SiO2 NPs on the Structural Properties of Fly Ash Based Sustainable Geopolymer. *Constr. Build. Mater.* **2018**, *164*, 238–245. [CrossRef]
14. DeMerlis, C.C.; Schoneker, D.R. Review of the Oral Toxicity of Poly Vinyl Alcohol (PVA). *Food Chem. Toxicol.* **2003**, *41*, 319–326. [CrossRef]
15. Muppalaneni, S. Polyvinyl Alcohol in Medicine and Pharmacy: A Perspective. *J. Dev. Drugs* **2013**, *2*, 2329–6631. [CrossRef]
16. PVOH Polymers LTD. Soluble Polymers For Innovation. Available online: https://www.pvohpolymers.co.uk/fisheries/ (accessed on 27 October 2021).
17. Porter, J.J.; Snider, E.H. Long Term Biodegradability of Textile Chemicals. *J. Water Pollut. Control Fed.* **1976**, *48*, 2198–2210.
18. Wu, H.F.; Yue, L.Z.; Jiang, S.L.; Lu, Y.Q.; Wu, Y.X.; Wan, Z.Y. Biodegradation of Polyvinyl Alcohol by Different Dominant Degrading Bacterial Strains in a Baffled Anaerobic Bioreactor. *Water Sci. Technol.* **2019**, *79*, 2005–2012. [CrossRef]
19. Watanabe, Y.; Morita, M.; Hamada, N.; Tsujisaca, Y. Formation of Hydrogen Peroxide by a Polyvinyl Degrading Alcohol Enzyme. *Agr. Biol.* **1975**, *39*, 2447–2448.
20. Suzuki, T.; Ichihara, Y.; Yamada, M.; Tonomura, K. Some Characteristics of Pseudomonas O–3 Which Utilizes Polyvinyl Alcohol. *Agric. Biol. Chem.* **1973**, *37*, 747–756. [CrossRef]
21. Chiellini, E.; Corti, A.; D'Antone, S.; Solaro, R. Biodegradation of Poly (Vinyl Alcohol) Based Materials. *Prog. Polym. Sci.* **2003**, *28*, 963–1404. [CrossRef]
22. Yamatsu, A.; Matsumi, R.; Atomi, H.; Imanaka, T. Isolation and Characterization of a Novel Poly (Vinyl Alcohol)-Degrading Bacterium, Sphingopyxis Sp. PVA3. *Appl. Microbiol. Biotechnol.* **2006**, *72*, 804–811. [CrossRef]
23. Rujnić Havstad, M.; Juroš, L.; Katančić, Z.; Pilipović, A. Influence of Home Composting on Tensile Properties of Commercial Biodegradable Plastic Films. *Polymers* **2021**, *13*, 2785. [CrossRef]
24. Lambert, S.; Wagner, M. Environmental Performance of Bio-Based and Biodegradable Plastics: The Road Ahead. *Chem. Soc. Rev.* **2017**, *46*, 6855–6871. [CrossRef]
25. Bagheri, A.R.; Laforsch, C.; Greiner, A.; Agarwal, S. Fate of So-Called Biodegradable Polymers in Seawater and Freshwater. *Glob. Chall.* **2017**, *1*, 1700048. [CrossRef]
26. Thellen, C.; Coyne, M.; Froio, D.; Auerbach, M.; Wirsen, C.; Ratto, J.A. A Processing, Characterization and Marine Biodegradation Study of Melt-Extruded Polyhydroxyalkanoate (PHA) Films. *J. Polym. Environ.* **2008**, *16*, 1–11. [CrossRef]
27. Accinelli, C.; Saccà, M.L.; Mencarelli, M.; Vicari, A. Deterioration of Bioplastic Carrier Bags in the Environment and Assessment of a New Recycling Alternative. *Chemosphere* **2012**, *89*, 136–143. [CrossRef]
28. Tachibana, K.; Urano, Y.; Numata, K. Biodegradability of Nylon 4 Film in a Marine Environment. *Polym. Degrad. Stab.* **2013**, *98*, 1847–1851. [CrossRef]
29. Tosin, M.; Weber, M.; Siotto, M.; Lott, C.; Innocenti, F.D. Laboratory Test Methods to Determine the Degradation of Plastics in Marine Environmental Conditions. *Front. Microbiol.* **2012**, *3*, 1–9. [CrossRef]
30. Beiras, R.; Durán, I.; Parra, S.; Urrutia, M.B.; Besada, V.; Bellas, J.; Viñas, L.; Sánchez-Marín, P.; González-Quijano, A.; Franco, M.A.; et al. Linking Chemical Contamination to Biological Effects in Coastal Pollution Monitoring. *Ecotoxicology* **2012**, *21*, 9–17. [CrossRef]
31. Cole, M.; Lindeque, P.; Fileman, E.; Halsband, C.; Galloway, T.S. The Impact of Polystyrene Microplastics on Feeding, Function and Fecundity in the Marine Copepod Calanus Helgolandicus. *Environ. Sci. Technol.* **2015**, *49*, 1130–1137. [CrossRef]
32. Burns, E.E.; Boxall, A.B.A. Microplastics in the Aquatic Environment: Evidence for or against Adverse Impacts and Major Knowledge Gaps. *Environ. Toxicol. Chem.* **2018**, *37*, 2776–2796. [CrossRef]
33. Cormier, B.; Gambardella, C.; Tato, T.; Perdriat, Q.; Costa, E.; Veclin, C.; Le Bihanic, F.; Grassl, B.; Dubocq, F.; Kärrman, A.; et al. Chemicals Sorbed to Environmental Microplastics Are Toxic to Early Life Stages of Aquatic Organisms. Ecotoxicol. *Environ. Saf.* **2021**, *208*, 111665. [CrossRef]
34. Beiras, R.; Verdejo, E.; Campoy-López, P.; Vidal-Liñán, L. Aquatic Toxicity of Chemically Defined Microplastics Can Be Explained by Functional Additives. *J. Hazard. Mater.* **2021**, *406*, 124338. [CrossRef]
35. Druege, U. Ethylene and Plant Responses to Abiotic Stress. In *Ethylene Action in Plants*; Springer: Berlin/Heidelberg, Germany, 2006; pp. 81–118.
36. González-Pleiter, M.; Tamayo-Belda, G.; Pulido-Reyes, G.A.; Leganés, F.; Rosal, R.; Fernández-Piñas, F. Secondary Nanoplastics Released from a Biodegradable Microplastic Severely Impact Freshwater Environments. *Environ. Sci. Process. Impacts* **2019**, *6*, 1382–1392. [CrossRef]
37. Tarique, J.; Sapuan, S.M.; Khalina, A. Effect of Glycerol Plasticizer Loading on the Physical, Mechanical, Thermal, and Barrier Properties of Arrowroot (Maranta Arundinacea) Starch Biopolymers. *Sci. Rep.* **2021**, *11*, 1–17. [CrossRef]
38. ISO (International Organization for Standardization). *ISO 14851: Determination of the Ultimate Aerobic Biodegradability of Plastic Materials in an Aqueous Medium—Method by Measuring the Oxygen Demand in a Closed Respirometer*; ISO: Geneva, Switzerland, 2019.
39. Redfield, A.C. The biological control of chemical factors in the environment. *Am. Sci.* **1958**, *46*, 205–221.
40. Tortell, P.D.; Maldonado, M.T.; Granger, J.; Price, N.M. Marine bacteria and biogeochemical cycling of iron in the oceans. *FEMS Microbiol. Ecol.* **1999**, *29*, 1–11. [CrossRef]
41. Beiras, R.; Tato, T.; López-Ibáñez, S. A 2-Tier standard method to test the toxicity of microplastics in marine water using Paracentrotus lividus and Acartia clausi larvae. *Environ Toxicol Chem.* **2018**, *38*, 630–637. [CrossRef]

42. Lorenzo, J.I.; Nieto, O.; Beiras, R. Effect of Humic Acids on Speciation and Toxicity of Copper To *Paracentrotus lividus* larvae in seawater. *Aquat. Toxicol.* **2002**, *58*, 27–41. [CrossRef]
43. Saco-Álvarez, L.; Durán, I.; Ignacio Lorenzo, J.; Beiras, R. Methodological Basis for the Optimization of a Marine Sea-Urchin Embryo Test (SET) for the Ecological Assessment of Coastal Water Quality. *Ecotoxicol. Environ. Saf.* **2010**, *73*, 491–499. [CrossRef]
44. (ISO) International Organization for Standardization. *ISO 19679: Determination of Aerobic Biodegradation of Non-Floating Plastic Materials in a Seawater/Sediment Interface—Method by Analysis of Evolved Carbon Dioxide*; ISO: Geneva, Switzerland, 2016.
45. ASTM (American Society for Testing and Materials). *D7991–15 Standard Test Method for Determining Aerobic Biodegradation of Plastics Buried in Sandy Marine Sediment under Controlled Laboratory Conditions*; ASTM: West Conshohocken, PA, USA, 2015.
46. Lin, C.C.; Lee, L.T.; Hsu, L.J. Degradation of Polyvinyl Alcohol in Aqueous Solutions Using UV-365 Nm/S2O82- Process. *Int. J. Environ. Sci. Technol.* **2014**, *11*, 831–838. [CrossRef]
47. Folino, A.; Karageorgiou, A.; Calabrò, P.S.; Komilis, D. Biodegradation of Wasted Bioplastics in Natural and Industrial Environments: A Review. *Sustainability* **2020**, *12*, 6030. [CrossRef]
48. Byrne, D.; Boeije, G.; Croft, I.; Hüttmann, G.; Luijkx, G.; Meier, F.; Parulekar, Y.; Stijntjes, G. Biodegradability of Polyvinyl Alcohol Based Film Used for Liquid Detergent Capsules. *Tenside Surfactants Deterg.* **2021**, *58*, 88–96. [CrossRef]
49. Abdullah, Z.W.; Dong, Y. Biodegradable and Water Resistant Poly(Vinyl) Alcohol (PVA)/Starch (ST)/Glycerol (GL)/Halloysite Nanotube (HNT) Nanocomposite Films for Sustainable Food Packaging. *Front. Mater.* **2019**, *6*, 1–17. [CrossRef]
50. Reuschenbach, P.; Pagga, U.; Strotmann, U. A Critical Comparison of Respirometric Biodegradation Tests Based on OECD 301 and Related Test Methods. *Water Res.* **2003**, *37*, 1571–1582. [CrossRef]
51. Wei, X.F.; Bohlén, M.; Lindblad, C.; Hedenqvist, M.; Hakonen, A. Microplastics generated from a biodegradable plastic in freshwater and seawater. *Water Res.* **2021**, *15*, 117123. [CrossRef]
52. TUV AUSTRIA. Certification Scheme OK Biodegradable Water. Available online: https://www.tuv-at.be/green-marks/doc-center/ (accessed on 17 October 2021).
53. Harrison, J.; Boardman, C.; O'Callaghan, K.; Delort, A.-M.; Song, J. Biodegradability standards for carrier bags and plastic films in aquatic environments: A critical review. *R. Soc. Open Sci.* **2018**, *5*, 171792. [CrossRef] [PubMed]
54. European Commission. Regulation (EC) No 1272/2008 of the European Parliament and of the Council of 16 December 2008 on Classification, Labelling and Packaging of Substances and Mixtures. *Off. J. Eur. Union* **2008**, 353.
55. Arfsten, D.P.; Burton, D.T.; Fisher, D.J.; Callahan, J.; Wilson, C.L.; Still, K.R.; Spargo, B.J. Assessment of the aquatic and terrestrial toxicity of five biodegradable polymers. *Environ. Res.* **2004**, *94*, 198–210. [CrossRef] [PubMed]
56. ECHA. Short-Term Toxicity to Aquatic Invertebrates. Available online: https://echa.europa.eu/es/registration-dossier/-/registered-dossier/14481/6/2/4 (accessed on 27 August 2021).
57. Perales, E.; García, C.B.; Lomba, L.; García, J.I.; Pires, E.; Sancho, M.C.; Navarro, E.; Giner, B. Comparative Ecotoxicity Study of Glycerol-Biobased Solvents. *Environ. Chem.* **2017**, *14*, 370–377. [CrossRef]
58. Gu, Y.; Jérôme, F. Glycerol as a sustainable solvent for green chemistry. *Green Chem.* **2010**, *12*, 1127–1138. [CrossRef]

polymers

MDPI

Article

Effect of Microcapsules with Waterborne Coating as Core Material on Properties of Coating for Tilia Europaea and Comparison with Other Microcapsules

Xiaoxing Yan [1,2,*], Yu Tao [2] and Xingyu Qian [2]

[1] Co-Innovation Center of Efficient Processing and Utilization of Forest Resources, Nanjing Forestry University, Nanjing 210037, China
[2] College of Furnishings and Industrial Design, Nanjing Forestry University, Nanjing 210037, China; taoyu@njfu.edu.cn (Y.T.); qianxingyu@njfu.edu.cn (X.Q.)
* Correspondence: yanxiaoxing@njfu.edu.cn

Abstract: Urea formaldehyde was used as wall material and waterborne coatings as a core material to prepare microcapsules. So as to explore the influence of mass ratio of core to shell, reaction temperature and standing time on the performance of microcapsules, the orthogonal test of three factors and two levels was put into effect. The orthogonal experimental results showed the mass ratio of core to shell was the most important factor. With the increase of the mass ratio of core to shell, the output and clad ratio of microcapsules increased first and then decreased. The microcapsule with the mass ratio of core to shell of 0.67:1 had better appearance, output, and encapsulation performance. The optical properties of waterborne wood coating with the microcapsules of waterborne coating as core materials did not decrease significantly, while the hardness, impact resistance, and toughness were improved. At the same time, the microcapsules have a certain self-repairing effect on coating micro-cracks. Compared with the properties of waterborne coatings with other microcapsules, the coating with waterborne coating as core material has better comprehensive performance. The results provide a new research idea for the performance enhancement and self-healing of wood waterborne coating.

Keywords: waterborne coating; microcapsule; self-healing; core material; clad ratio

Citation: Yan, X.; Tao, Y.; Qian, X. Effect of Microcapsules with Waterborne Coating as Core Material on Properties of Coating for Tilia Europaea and Comparison with Other Microcapsules. *Polymers* **2021**, *13*, 3167. https://doi.org/10.3390/polym13183167

Academic Editors: Florian J. Stadler and Alberto García-Peñas

Received: 13 July 2021
Accepted: 12 September 2021
Published: 18 September 2021

1. Introduction

In the process of using the coating, owing to the changes of environmental factors (such as light, temperature, humidity), the low toughness, the poor adaptability, and the complex process, the micro-cracks are generated and expanded inside the coating, and even destroy the internal structure of the coating, thus influencing the operation life of the coating and wood [1–3]. Because of the unique wet swelling and drying shrinkage and dimensional instability of wood [4], the microcapsule self-healing technology has a broad prospect in wood materials [5,6]. When the coating [7] is cracked by the external force, the microcapsule is cracked and the core material healing agent is released [8], which can enhance the performance of wood and prolong its service life. Waterborne acrylic coating is a commonly used waterborne coating [9,10]. It uses water as the dispersion medium, and does not contain formaldehyde and heavy metals [11,12]. It is non-toxic and has no pungent smell, and its harm to human health and environmental pollution is relatively small, so it can meet the increasingly stringent environmental protection requirements [13–15].

At present, the microcapsules have been widely used in wood materials [16,17]. In the combination of microcapsules and wood materials, Jeong et al. [18] used microencapsulated phase change materials to increase its heat storage, and introduced them into wood-based panels by physical mixing of microcapsules and adhesives, in order to produce high thermal efficiency wood flooring. Pinkl et al. [19] further proposed that if the self-healing material is added to repair the damaged adhesive joint, the adhesive closer to the bonding line may be better. He et al. [20] used microwave expansion and other pretreatment technologies to increase the gap of wood to achieve the purpose of introducing larger diameter microcapsules. For the application of microcapsules in wood functional materials [21], it is mainly explored how to introduce microcapsules into wood or wood materials. However, this combination process is more complex. According to the requirements of wood outputs for single or composite functional characteristics, microcapsule are evenly introduced into wood and wood materials, or attached to the surface of wood and wood materials by surface modification technology [22–24]. In the application of microcapsules to self-healing coatings [25], Siva et al. [26] used linseed oil and mercapto benzothiazole as core materials, and compounded urea formaldehyde microcapsules by emulsion polymerization. The self-healing ability of the anticorrosive coating with microcapsules was discussed by SVET. Lang et al. [27] encapsulated linseed oil in polyurea formaldehyde (PUF) shell and prepared a self-healing coating on account of microcapsules successfully. Meanwhile, the influence of the amount of microcapsules on the repair property of the paint-coat was explored. Yang et al. [28] explored a new self-healing system based on microcapsule organogels, where there would be no damage in the healing area. The self-healing coating was prepared by microencapsulation of urea formaldehyde polymer into polymer coating. As for the application of microcapsule in self-healing coating, scholars mainly explored its role in metal surface anti-corrosion coating, and often used linseed oil and other dry oil as core material repair agent, while there are few studies on the application of microcapsules in coatings of wood surface [29]. In order to simplify the combination of wood and self-healing microcapsule, the self-healing microcapsule with waterborne coating as the core material is applied to the coating on the surface of wood in this paper. It is of great significance to apply self-healing microcapsules to prepare self-healing coatings for wood products. The core material of self-healing microcapsules has an important influence on the properties of microcapsules. If the core material repair agent and coating are two different substances, after the coating has micro cracks, when the core material repair agent is curing and healing micro cracks, the combination of the repair agent and coating may produce interfacial stress, which affects the repair effect of micro cracks. Therefore, the choice of microcapsule core material will not only affect the performance of the microcapsule itself, but also have a great impact on the self-healing performance of the coating. As the wall material of microcapsules, urea formaldehyde resin has the advantages of low cost, mature process, and good coating of core materials. However, free formaldehyde gas may be released in the preparation process, causing harm to the human body. Therefore, it is necessary to control the content of formaldehyde and reduce the release of formaldehyde.

The performance of waterborne coating is better than that of single resin because it contains resin, additive, and other compounds. Using waterborne acrylic coating as the core material, the repairing agent can be cured at room temperature to form a coating after micro cracks [30,31]. Because the composition of the core material is consistent with that of the waterborne coating itself, the interface compatibility will be better when the crack is healed, and the performance of the coating can be further improved. In order to further explore the effect of the core material, this paper used waterborne coating itself as the core material repair agent to prepare microcapsules, and its process parameters were optimized. Then, the prepared microcapsules were added to waterborne coating to explore the effect of the self-healing microcapsules on the coating properties. On this basis, the coating properties of microcapsules were compared with those of other microcapsules with different core materials reported in the literature [32–34]. The main objective of the research is to explore the influence of different core materials on the coating and obtain better self-healing effect, so as to provide the basis for the application of self-healing microcapsules in wood materials in the future.

2. Materials and Methods

2.1. Materials

The chemical reagent information required for the experiment is shown in Table 1. Nippon waterborne acrylic coatings, which are mainly composed of waterborne acrylic copolymer, additives, and water, was provided by Nippon (Changzhou) Co., Ltd., Changzhou, China. Tilia europaea with size of 100 × 50 × 4 and uniform color were provided by Guangdong Yihua life science and Technology Co., Ltd., Shantou, China.

Table 1. Chemical reagent information in experiment.

Experimental Materials	Molecular Mass (g/moL)	CAS	Manufacturer
citric acid monohydrate	210.14	5949-29-1	Nanjing Chemical Reagent Co., Ltd., Nanjing, China
triethanolamine	149.19	102-71-6	Guangzhou Jiangshun Chemical Technology Co., Ltd., Guangzhou, China
37.0% formaldehyde	30.03	50-00-0	Guangzhou Jiangshun Chemical Technology Co., Ltd., Guangzhou, China
urea	60.06	57-13-6	Nanjing Chemical Reagent Co., Ltd., Nanjing, China
sodium dodecyl benzene sulfonate (SDBS)	348.48	25155-30-0	Xi'an Tianmao Chemical Co., Ltd., Xi'an, China
ethyl acetate	88.11	141-78-6	Tianjin Fuyu Fine Chemical Co., Ltd., Tianjin, China
anhydrous ethanol	46.07	64-17-5	Tianjin Fuyu Fine Chemical Co., Ltd., Tianjin, China

2.2. Preparation of Microcapsules

Firstly, according to the related literature of microcapsules [35], three vital factors affecting the preparation of microcapsules were determined. As shown in Table 2, the orthogonal test of 3 factors and 2 levels was conducted. Two levels of mass ratio of core to shell, reaction temperature, and standing time were chosen to prepare urea formaldehyde-coated waterborne coating microcapsules.

Table 2. Influencing factors and levels.

Level	Mass Ratio of Core to Shell	Reaction Temperature (°C)	Standing Time (d)
1	0.42:1	50	1
2	0.67:1	70	5

Samples 1–4# microcapsules of the orthogonal test were prepared, and the details of each substance dosage were listed in Table 3. Taking sample 1# of orthogonal experiment as an example, the preparation process of microcapsules was explained. The formation process of microcapsules is shown in Figure 1. The first step was the synthesis of urea formaldehyde (UF) prepolymer: after 20.0 g urea was poured into the beaker, 27.0 g of 37.0% formaldehyde solution was added into it and fully stirred. Then, the triethanolamine was put in slowly to regulate the pH to 8.5–9.0, and the mixture was continuously stirred at the rate of 100 rpm in a stationary temperature water bath at 70 °C for 30 min to obtain a faintly thick and transparent UF prepolymer solution. The second step was the emulsification of waterborne coatings for core material: 0.975 g of SDBS was put into distilled water and stirred thoroughly to obtain a mixture with concentration of 1.0% as emulsifier. Next the 12.5 g waterborne coating was mixed with 97.0 mL emulsifier solution. The mixture was emulsified for 60 min at the rate of 1200 rpm and 50 °C reaction temperature to gain steady emulsion. The last step was microencapsulation: the wall material prepolymer was slowly dropped into the core material emulsion obtained in the second step at the speed of 300 rpm, then citric acid solution was dropped gradually, and the mixture was continuously stirred until the citric acid crystal was completely dissolved. After adjusting the pH value to 3.0–3.5, the temperature was tardily increased to 50 °C and kept for 2.5 h.

After 1 d of standing time, the production was filtered and distilled water was used at the same time to wash off the surplus emulsifier. Then, the production was laid into the drying oven and heated at 45 °C for 24 h. The powder gained was the required sample 1# in the orthogonal experiment. The preparation process of other microcapsules samples is consistent with the above.

Table 3. Detailed list of various substances used in orthogonal and single factor tests.

Sample	37.0% Formaldehyde (g)	Urea (g)	Waterborne Coatings (g)	Sodium Dodecyl Benzene Sulfonate (g)	Emulsifier (mL)
1#	27.00	20.00	12.50	0.97	97.00
2#	27.00	20.00	12.50	0.97	97.00
3#	27.00	20.00	20.00	1.56	156.00
4#	27.00	20.00	20.00	1.56	156.00
5#	27.00	20.00	12.50	0.97	97.00
6#	27.00	20.00	15.00	1.17	117.00
7#	27.00	20.00	17.50	1.37	137.00
8#	27.00	20.00	20.00	1.56	156.00
9#	27.00	20.00	22.50	1.76	176.00
10#	27.00	20.00	25.00	1.95	195.00
11#	27.00	20.00	27.50	2.15	215.00

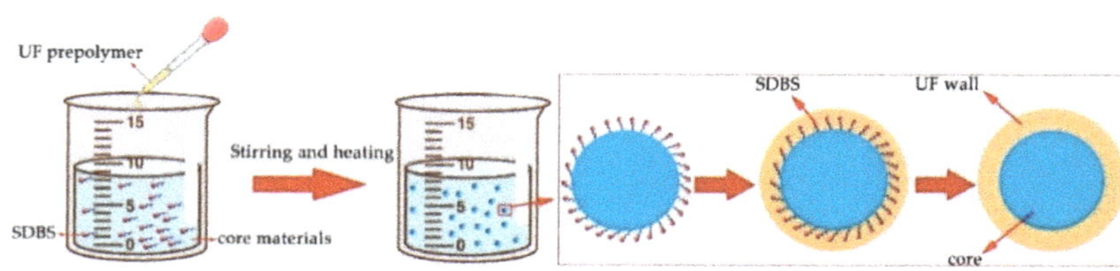

Figure 1. The formation process of microcapsules.

Through the above orthogonal experiment, the most influential factor of the three factors on the performance of microcapsules and the better preparation scheme of microcapsules were determined, and next the independent test was developed for the most influential element. The experimental materials, equipment, and preparation method of microcapsules remained unchanged. The reaction temperature was regulated to 70 °C, and the standing time was set at 5 d. As an independent variable, the mass ratio of core to shell was 0.42:1, 0.50:1, 0.58:1, 0.67:1, 0.75:1, 0.83:1, and 0.92:1 (samples 5–11#), respectively. The consumption details of each substance are shown in Table 3.

2.3. Preparation of Coating with Microcapsules

According to the relevant literature, when the mass fraction of epoxy resin microcapsule, shellac microcapsule, and waterborne acrylic acid microcapsule in waterborne coating is 10.0%, the coating performance is better [32–34]. Therefore, in order to carry out the subsequent comparative analysis, the optimized mass ratio of core to shell of 0.67:1 waterborne coating microcapsules were added to the waterborne coating with a mass fraction of 0 and 10.0%, respectively, and stirred evenly. The prepared coating was applied on Tilia europaea substrate to form a uniform coating on its surface. After the coating was completely dry, 800 mesh sandpaper [36] was used to grind the paint coating to make it smooth and the surface was mopped up with dry cloth. The above procedure was performed three times in the same way to obtain a coating thickness of about 60 μm.

2.4. Testing and Characterization

In the aspect of micro morphology of microcapsules, a Zeiss Axio scope A1 optical microscope (OM) and a FEI Quanta−200 scanning electron microscope (SEM), produced by FEI Company, Hillsboro, OR, USA, were employed for characterization. In terms of chemical composition, A VERTEX 80 V Fourier infrared spectrum analyzer (FTIR), produced by Bruker (Beijing) Scientific Technology Co., Ltd., Beijing, China, was applied for analysis. For the microcapsule clad ratio test, the mass of microcapsule (m_1) was fully ground and soaked in ethyl acetate for 72 h. The solution should be changed every 24 h. Finally, the filtered product was washed with deionized water, and the mass of the residue (m_2) after drying was weighed. The encapsulation rate (c) of microcapsules was figured out by Formula (1).

$$c = \frac{m_1 - m_2}{m_1} \times 100\% \tag{1}$$

In the aspect of coating optical performance, the chromaticity value of waterborne coating on Tilia europaea surface was measured by RM200-PT01 color detector (Pantone Inc., Carlstadt, NJ, USA). The HG268 trigonometric glossmeter (Shenzhen Forbes Instrument Co., Ltd., Shenzhen, China) was applied to test the gloss of waterborne coating film, and the gloss of coating at 60° incidence angle was recorded. As for the mechanical performance of the coating, the pencil with known hardness label was applied to characterize the hardness of the coating. In the process, the hardness of the pencil was gradually increased until the surface of the coating appeared defects such as plastic surface or cohesive damage. The pencil hardness of the coating was expressed by the hardest hardness that does not cause a scratch of 3 mm or more. The QFH-HG600 film scribbler (Guangzhou Yuanxiao Marine Equipment Co., Ltd., Guangzhou, China) was operated to characterize the adhesion of the coating, and the resistance of the coating falling off from the substrate was evaluated when the coating penetrated the substrate with a right angle grid pattern. According to the surface appearance of the cross cutting area, it can be divided into six grades: 0, 1, 2, 3, 4, and 5, and the first three grades were satisfactory. The QCJ film impact test instrument (Tianjin World Expo Weiye chemical Glass Instrument Co., Ltd., Tianjin, China) was applied to test the coating impact resistance. A 1.0 kg heavy hammer was fixed at different heights that were less than 50 cm, and next the heavy hammer was allowed to fall naturally to impact the test piece to observe whether cracks and spalling phenomena appeared on the surface of coating. Finally, when there was no crack in the coating, the maximum height of the hammer was the impact strength of the coating. The tensile strength of the coating was tested by AG-IC100KN universal testing machine (Guangzhou Kehui Instrument Co., Ltd., Guangzhou, China). Then, the elongation at break of the coating was counted by the tensile strength. All the tests were repeated 4 times, and the error was less than 5.0%.

3. Results and Discussion

3.1. *Analysis of Orthogonal Test Results*

3.1.1. Analysis of Micro Characteristics of Microcapsules

The macroscopic images of microcapsule powder are shown in Figure 2. The SEM appearance of the orthogonal samples 1–4# in Table 3 was displayed in Figure 3. The morphology of sample 4# microcapsule (Figure 3D) is relatively good, followed by sample 1# (Figure 3A). Sample 1# has a general coating condition, with some agglomeration, but the particle size is basically uniform, about 5–8 μm. Sample 2# (Figure 3B) has a poor coating condition, with more urea formaldehyde precipitation and agglomeration, and the particle size is about 8–10 μm. Sample 3# (Figure 3C) has a general coating, with less precipitation and agglomeration, and its particle size is about 5–8 μm. Sample 4# microcapsule is coated well with a little agglomeration, and the particle size is about 8 μm. It can be found that the UF resin coated waterborne coating microcapsules whose particle size was approximately 8 μm have been successfully prepared.

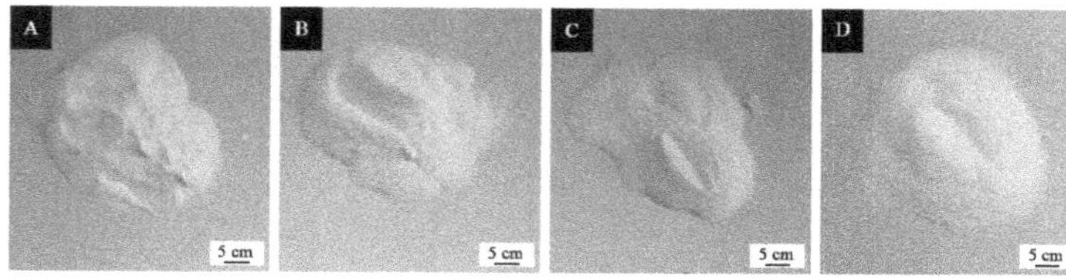

Figure 2. The macroscopic images of microcapsule powder: (**A**) sample 1#, (**B**) sample 2#, (**C**) sample 3#, (**D**) sample 4#.

Figure 3. The SEM morphologies of microcapsules under different orthogonal experimental parameters: (**A**) sample 1#, (**B**) sample 2#, (**C**) sample 3#, (**D**) sample 4#.

3.1.2. Infrared Analysis of Components in Microcapsules

Figure 4 shows the FTIR spectrogram of wall material, core material and orthogonal test sample 4#. The special absorption peaks at 3350 and 1556 cm^{-1} are assigned to N-H and C-N functional groups, which belong to the functional groups of UF resin. The absorption at 1641 cm^{-1} is consistent with the stretching vibration of C=O functional group in UF resin. The absorptions at 2961 and 1447 cm^{-1} are the special absorption peaks of C-H. The special absorption at 1726 cm^{-1} belongs to the characteristic absorption of C=O functional group in the waterborne acrylic, and the corresponding peak also appears in the infrared spectrum of sample 4# in the orthogonal test. It is confirmed that the corresponding UF resin and waterborne acrylic exist in the prepared microcapsule, and the composition is not damaged. It proves the successful preparation of UF resin cladding waterborne coating microcapsules.

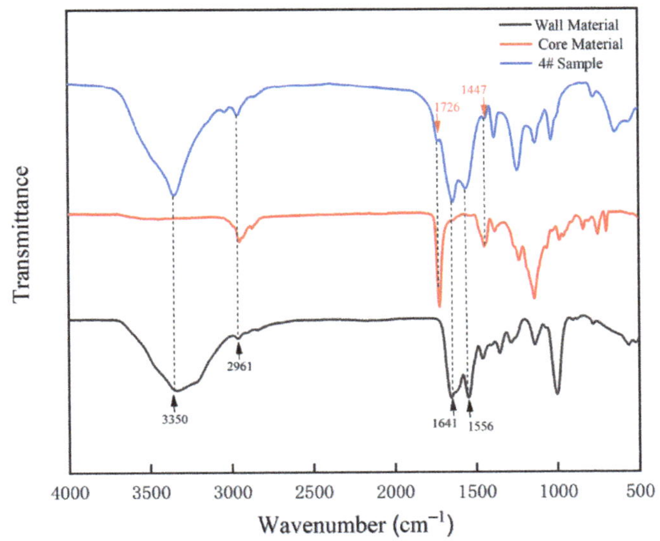

Figure 4. The FTIR spectrogram of wall material, core material, and sample 4# microcapsule.

3.1.3. Output Analysis of Microcapsules

The microcapsule samples 1–4# made by orthogonal experiment were weighed, and the range (R) and variance results of output are shown in Tables 4 and 5. When the mass ratio of core to wall and the standing time increased, the output of the microcapsule increased, and when the reaction temperature increased, the output of the microcapsule decreased. It can be found from the Table 4 that sample 3# microcapsule has the largest output at 33.32 g. From the range results, it can be seen that in the three factors of the mass ratio of core to shell, reaction temperature, and standing time, the mass ratio of core to shell makes the greatest influence on the output of urea formaldehyde-coated waterborne coating microcapsules, followed by the standing time. Only from the results of output, the better preparation parameters of the microcapsule are mass ratio of core to shell of 0.67:1, the reaction temperature of 50 °C and the standing time of 5 d. In Table 5, F ratio was obtained by comparing the square sum of the average deviation of each factor with the square sum of the average deviation of the error. This ratio reflected the influence of each factor on the test results. If F ratio was greater than the F critical value, it showed that this factor had a significant influence on the test results. It can be seen from Table 5 that the F ratio of the mass ratio of core to shell is greater than the F critical value, indicating that the mass ratio of core to shell has a significant impact on the output of microcapsules.

Table 4. Range results of microcapsule output by the orthogonal test.

Sample	Mass Ratio of Core to Shell	Reaction Temperature (°C)	Standing Time (d)	Output (g)
1#	0.42:1	50	1	28.76 ± 0.95
2#	0.42:1	70	5	29.34 ± 1.02
3#	0.67:1	50	5	33.32 ± 0.84
4#	0.67:1	70	1	32.19 ± 0.92
Mean 1	29.05 ± 1.38	31.04 ± 0.97	30.47 ± 0.78	
Mean 2	32.75 ± 0.85	30.76 ± 0.99	31.33 ± 0.83	
R	3.70 ± 0.10	0.27 ± 0.01	0.85 ± 0.01	

Table 5. Variance results of microcapsule output by the orthogonal test.

Factor	Sum of Squared Deviations	Degrees of Freedom	F Ratio [1]	F Critical Value [2]	Significance
Mass ratio of core to shell	13.727	1	180.618	161.000	*
Reaction temperature (°C)	0.076	1	1.000	161.000	
Standing time (d)	0.731	1	9.618	161.000	
Error	0.08	1			

[1] F ratio is obtained by comparing the square sum of the average deviation of each factor with the square sum of the average deviation of the error. [2] F critical value is the critical value of given significant level.

3.1.4. Clad Ratio Analysis of Microcapsules

The range and variance of clad ratio of UF resin cladding waterborne coating microcapsules made by orthogonal test are displayed in Tables 6 and 7. When the mass ratio of core to wall, the reaction temperature, and the standing time increased, the clad ratio of the microcapsule increased. Sample 3# of the microcapsule has the highest clad ratio of 43.0%. On the basis of the variance results of the clad ratio, it is obvious that the mass ratio of core to shell has the greatest influence on the clad ratio of the microcapsules, followed by the standing time. Based on the results of clad ratio, it can be found that the mass ratio of core to shell of 0.67:1, the reaction temperature of 70 °C and the standing time of 5 d are the optimum preparation process of the microcapsules.

Table 6. Range results of microcapsule clad ratio by the orthogonal test.

Sample	Mass Ratio of Core to Shell	Reaction Temperature (°C)	Standing Time (d)	Clad Ratio (%)
1#	0.42:1	50	1	33.0 ± 1.0
2#	0.42:1	70	5	37.0 ± 1.0
3#	0.67:1	50	5	43.0 ± 1.4
4#	0.67:1	70	1	40.0 ± 0.8
Mean 1	35.00 ± 1.01	38.00 ± 1.10	36.50 ± 0.90	
Mean 2	41.50 ± 0.76	38.50 ± 1.00	40.00 ± 1.18	
R	6.50 ± 0.18	0.50 ± 0.02	3.50 ± 0.06	

Table 7. Variance results of microcapsule clad ratio by the orthogonal test.

Factor	Sum of Squared Deviations	Degrees of Freedom	F Ratio	F Critical Value	Significance
Mass ratio of core to shell	42.25	1	160.00	161.00	
Reaction temperature (°C)	0.25	1	1.00	161.00	
Standing time (d)	12.25	1	49.00	161.00	
Error	0.25	1			

The orthogonal test results for this kind of microcapsule were comprehensively analyzed. Among the three elements, the mass ratio of core to shell is the most important element affecting the output and clad ratio results of microcapsule. In order to further optimize the appearance and optical and mechanical properties of microcapsules, a single factor test was carried out for the most significant influencing factor (mass ratio of core to shell) according to the above orthogonal experimental results. Based on the analysis of output results, it can be determined that the best levels of the two other factors are reaction temperature of 50 °C and standing time of 5 d, respectively, and the best levels of the analysis of clad ratio results are 70 °C and 5 d. Compared with the output, the clad ratio is a more important factor in judging the performance of microcapsules. Because the micro morphology of sample 4# microcapsule was the best, the reaction temperature was regulated to 70 °C in the process of preparing sample 4#, finally the reaction temperature was adjusted to 70 °C and the standing time was set at 5 d in the single factor experiment.

3.2. Analysis of Single Factor Test Results

3.2.1. Analysis of Micro Characteristics of Microcapsules

The OM and SEM of microcapsules optimized by single factor experiment with mass ratio of core to shell are displayed in Figures 5 and 6. Figure 5H is the magnification image of Figure 5D. It can be seen from Figure 5H that the outside of the microcapsule sphere is the dark part, which is the wall material, and the bright spot of the microcapsule is the core material. The figure shows that the microcapsules with small mass ratio of core to shell of 0.42:1, 0.50:1, 0.58:1, and 0.67:1 (Figure 6A–D) have great morphology, consistent particle size, and almost no precipitation. However, when the mass ratio of core to shell increases, the content of core material increases, and the microcapsules with mass ratio of core to shells of 0.75:1, 0.83:1, and 0.92:1 (Figure 6E–G) can be found to have serious agglomeration and precipitation. The microcapsule with mass ratio of core to shell of 0.67:1 (Figures 5D and 6D) has the best micro morphology, and the particle size is 5 μm.

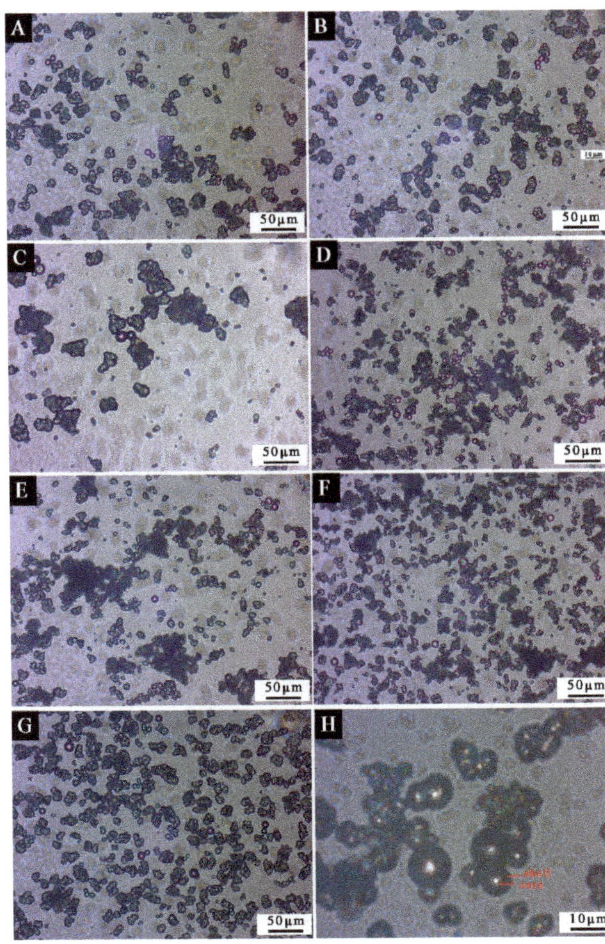

Figure 5. The OM appearance of microcapsules by single factor experiment: mass ratio of core to shell (**A**) 0.42:1, (**B**) 0.50:1, (**C**) 0.58:1, (**D**) 0.67:1, (**E**) 0.75:1, (**F**) 0.83:1, (**G**) 0.92:1, (**H**) 0.67:1.

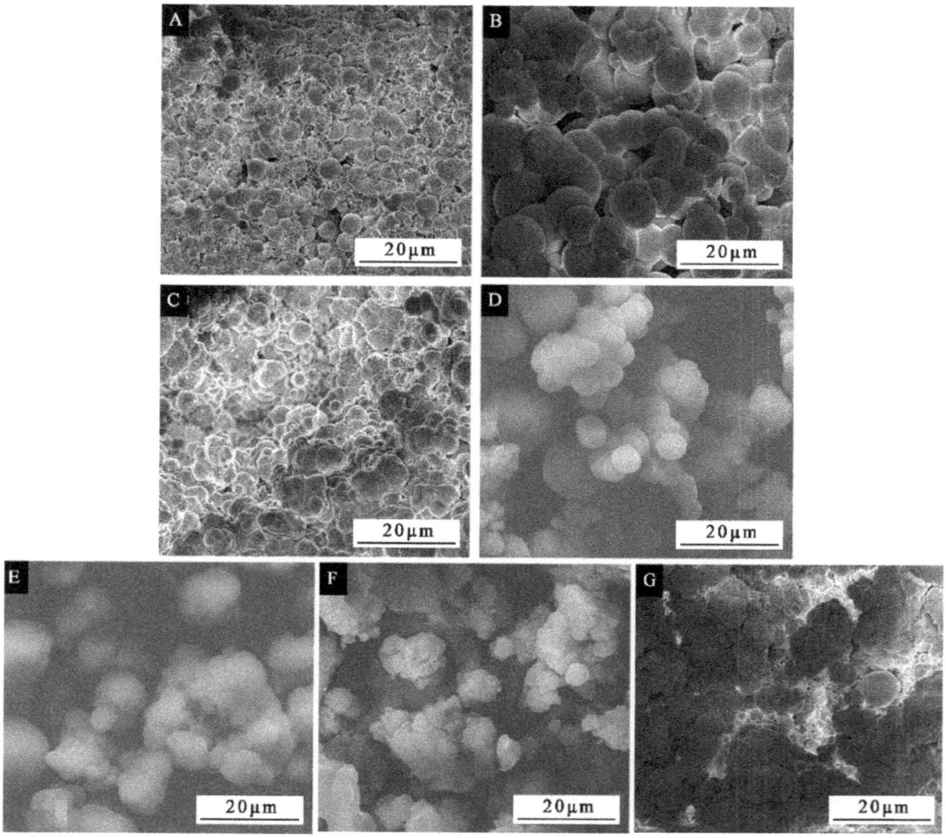

Figure 6. The SEM appearance of microcapsules by single factor experiment: mass ratio of core to shell (**A**) 0.42:1, (**B**) 0.50:1, (**C**) 0.58:1, (**D**) 0.67:1, (**E**) 0.75:1, (**F**) 0.83:1, (**G**) 0.92:1.

3.2.2. Infrared Analysis of Components in Microcapsules

Figure 7 shows the FTIR of seven kinds of microcapsules prepared by single factor experiment. The special absorption peaks of N-H and C-N functional groups are at 3350 and 1556 cm^{-1}. The absorption at 1639 cm^{-1} is the stretching vibration of C=O functional group in UF resin. The absorption at 2966 cm^{-1} is the special absorption peak of C-H functional group, and the peak at 1726 cm^{-1} represents the special absorption of C=O functional group in waterborne acrylic acid coatings. It can be found from the picture that the special absorption peaks of microcapsules by different ratios of core to wall are similar, which can prove that the chemical component part of microcapsules by different ratios of core to wall is still present, and the microcapsules are successfully prepared.

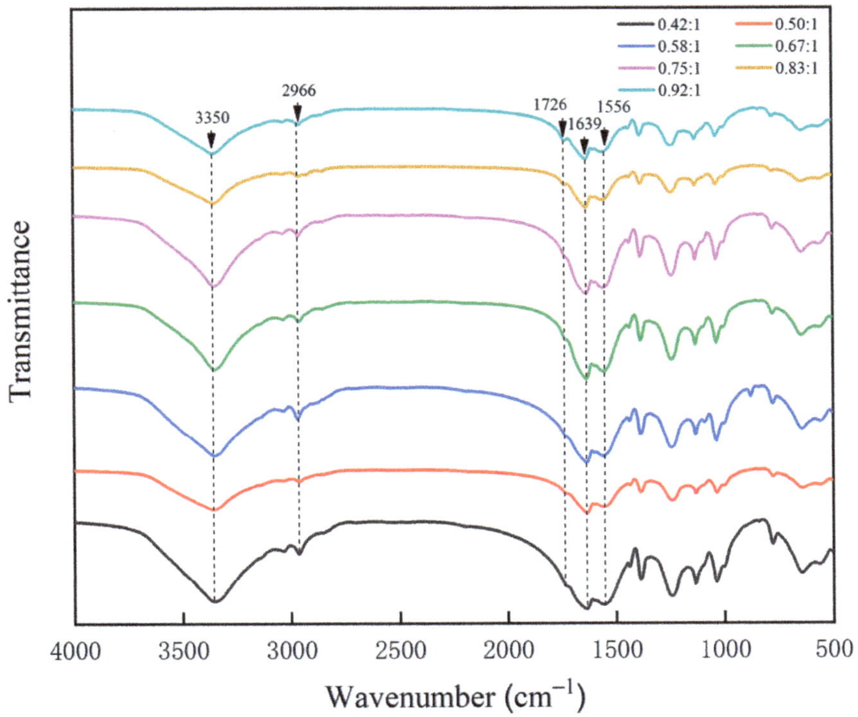

Figure 7. FTIR spectrum of microcapsules by single factor experiment.

3.2.3. Output Analysis of Microcapsules

The output of microcapsules with seven ratios of core to shell of waterborne coatings as core material is displayed in Table 8. The output results displayed that the microcapsule with mass ratio of core to shell of 0.67:1 had the highest output. As can be seen from the table, the overall output of microcapsules basically increased first and then decreased, with the increasing of mass ratio of core to shell. When the mass ratio of core to shell increased from 0.42:1 to 0.67:1, the output of microcapsules increased from 28.93 g to 37.80 g. The reason may be that the increase of the weight of the core material can increase the weight of the wall material to cover the repair agent, thus increasing the output of the microcapsule. However, as the mass ratio of core to shell continued to increase, the output decreased. The reason may be that the increase of core material led to the increase of agglomeration and precipitation. As a result, the excess core material cannot be completely covered and the mass loss occurred during the filtration process.

Table 8. Output results of microcapsules by the single factor test.

Sample	Mass Ratio of Core to Shell	Core Material Mass (g)	Output (g)
5#	0.42:1	12.5	28.93 ± 0.40
6#	0.50:1	15.0	31.22 ± 1.19
7#	0.58:1	17.5	34.35 ± 1.04
8#	0.67:1	20.0	37.80 ± 0.88
9#	0.75:1	22.5	35.57 ± 0.76
10#	0.83:1	25.0	33.43 ± 0.69
11#	0.92:1	27.5	33.47 ± 0.76

3.2.4. Clad Ratio Analysis of Microcapsules

The results of clad ratio of microcapsules with seven different mass ratios of core to shell in single factor experiment are shown in Table 9. From the results in the Table 9, when the mass ratio of core to shell was 0.67:1, the clad ratio of microcapsules was the highest. With the enlargement in the mass ratio of core to shell (the raise of the mass of core material), the clad ratio of waterborne coating microcapsules basically rose first then fell. When the mass ratio of core to shell increased from 0.42:1 to 0.67:1, the clad ratio of microcapsules rose from 37.0 to 49.0%, but when the mass ratio of core to shell rose to 0.92:1, the clad ratio of microcapsules fell to 32.0%. Owing to the raise of the core material mass, the content of the core material cladded in the microcapsule wall increased, which enhanced the clad ratio to a certain degree. Nevertheless, when the mass ratio of core to shell rose, the microcapsule would produce too much precipitation and agglomeration, which affected the performance of the microcapsule, resulting in the decrease of the clad ratio.

Table 9. Clad ratio results of microcapsules by the single factor test.

Sample	Mass Ratio of Core to Shell	Core Material Mass (g)	Clad Ratio (%)
5#	0.42:1	12.5	37.0 ± 1.2
6#	0.50:1	15.0	46.0 ± 1.4
7#	0.58:1	17.5	45.0 ± 1.3
8#	0.67:1	20.0	49.0 ± 1.8
9#	0.75:1	22.5	42.0 ± 1.2
10#	0.83:1	25.0	41.0 ± 1.3
11#	0.92:1	27.5	32.0 ± 0.6

Comprehensive analysis of the results in Figure 6, Tables 8 and 9 showed that when the mass ratio of core to shell was 0.67:1, the microcapsule had better morphology, the highest output, and the highest clad ratio, and the preparation was more successful.

3.3. *Effect of Standing Time on the Morphology of Microcapsules*

In order to understand the formation mechanism of microcapsules and explore the effect of standing time on the morphology of microcapsules, the microcapsules with mass ratio of core to shell of 0.67:1 were aged for 0, 1, 2, 3, 4, and 5 d respectively. The OM of the microcapsules prepared under different standing times is shown in Figure 8. With the increase of standing time, the observed spherical particles gradually increased, and with the extension of time, the particles formed a partial agglomeration phenomenon. In order to explore whether the agglomeration phenomenon of microcapsules would seriously increase with the infinite extension of time, the prepared microcapsules were aged for two months. The SEM under different multiples is shown in Figure 9. The microcapsules had good morphology and a small amount of agglomeration and bonding. The result showed that although a small amount of agglomeration occurred with the increase of standing time, it was not serious due to the long standing time. It also showed that the standing time had little effect, and the core wall ratio was a more important factor affecting the morphology of microcapsules.

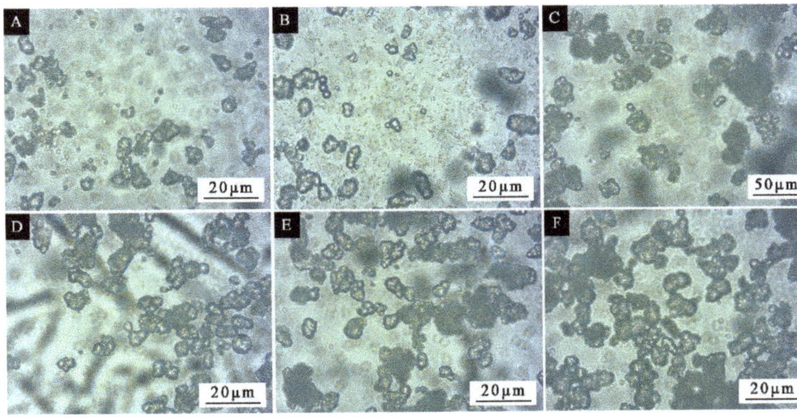

Figure 8. OM morphologies of microcapsules prepared under different depositing time: (**A**) 0, (**B**) 1 d, (**C**) 2 d, (**D**) 3 d, (**E**) 4 d, (**F**) 5 d.

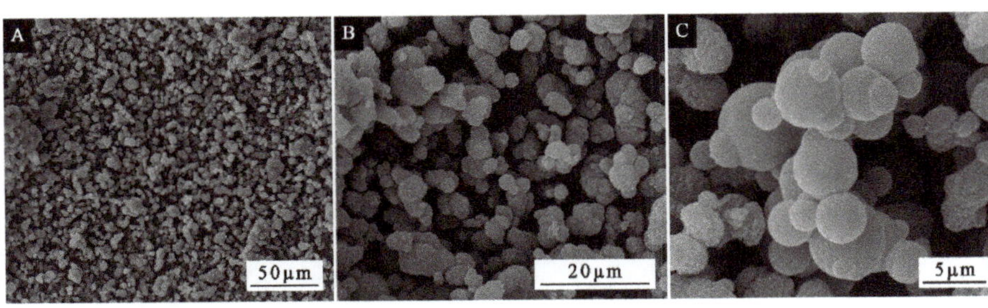

Figure 9. SEM morphology of microcapsule prepared under two months of deposition: (**A**) low magnification, (**B**) middle magnification, (**C**) large magnification.

3.4. Influence of Microcapsules on the Performances of Waterborne Coatings

The performance of the coating is shown in Table 10. Properties of coating with 10.0% mass fraction microcapsules were compared with that of the coating without microcapsule. In terms of optical properties, the color difference of the coating increased from 0.6 to 2.0 after adding 10.0% waterborne coating microcapsules, and the gloss of the coating fell from 29.4 to 11.9% at 60° incident angle. It may be due to the uneven surface of the coating and the enhancement of diffuse reflection after adding microcapsule particles, and the influence of the color of the microcapsule itself. In terms of mechanical properties, the hardness of the coating increased from HB to 3H, the adhesion remained unchanged, and the impact resistance raised significantly from 6.0 to 13.0 kg·cm. After comprehensive analysis, it can be concluded that the performance of the waterborne coating added with 10.0% waterborne coating microcapsules did not decrease significantly, but its hardness and impact resistance were enhanced. At the same time, the elongation at break of the coating had risen from 9.94 to 16.18%, which was a significant increase. It is obvious that the toughness of the coating with microcapsule concentration of 10.0% is improved, and the existence of microcapsule can prevent the appearance of micro-cracks to a certain extent.

Table 10. Performance comparison of waterborne coatings with different microcapsules.

Core Material	Optimum Mass Ratio of Core to Shell	Microcapsules Concentration (%)	Color Difference	Gloss (%)	Hardness	Adhesion (Grade)	Impact Resistance (kg·cm)	Elongation at Break (%)
–	–	0	0.60 ± 0.01	29.4 ± 0.8	HB ± 0	0 ± 0	6.0 ± 0.1	9.9 ± 0.2
Waterborne coating	0.67	10.0	2.00 ± 0.08	11.9 ± 0.3	3H ± 0	0 ± 0	13.0 ± 0.4	16.2 ± 0.5
Waterborne acrylic acid	0.58:1	10.0	1.80 ± 0.03	5.1 ± 0.1	2H ± 0	1 ± 0	15.0 ± 0.3	16.7 ± 0.5
Epoxy resin	0.83:1	10.0	3.50 ± 0.03	5.0 ± 0.1	5H ± 0	3 ± 0	20.0 ± 0.3	35.0 ± 1.0
Shellac [1]	0.75:1	10.0	1.50 ± 0.04	7.8 ± 0.2	B ± 0	1 ± 0	9.0 ± 0.2	20.9 ± 0.1

[1] Shellac is a kind of purple natural resin secreted by shellac insect after absorbing the sap of host tree.

In terms of self-healing performance, the blade was used to mark the cracks on the paint coating on the surface of Tilia europaea, and the OM was used to record the cracks. After an interval of 5 d, the OM was used to observe the self-healing effect of the paint coating again. According to Figure 10, the scratch width of the coating without microcapsules changed from 22.85 to 20.09 μm, and there was no obvious change. However, it can be seen from Figure 11 that the scratch width of the coating with 10.0% microcapsules decreased significantly from 28.72 to 20.71 μm after 5 d. It can be seen that the self-healing effect of the coating with microcapsule is better than that without microcapsule. Therefore, 10.0% microcapsules have a certain self-repairing effect on the micro cracks on the surface coating of wood. The self-healing mechanism of Tilia europaea surface coating containing microcapsules is shown in Figure 12. When the coating cracks, the microcapsules in the coating will break, and the waterborne coating core repair agent in the microcapsules will flow out. The waterborne paint repair agent can be cured into a film at room temperature. The curing of waterborne coating mainly has three levels of mechanisms. The first is the volatilization of water and other film-forming aids to fully reflect the basic properties of the thermoplastic resin itself, the second is the aggregation and fusion of the emulsion particles, and the third is mutual diffusion of particles. When the temperature is higher than the glass transition temperature, the final film is formed. The micro-cracks on the coating surface can achieve a certain self-repairing effect through the curing of the core material repairing agent.

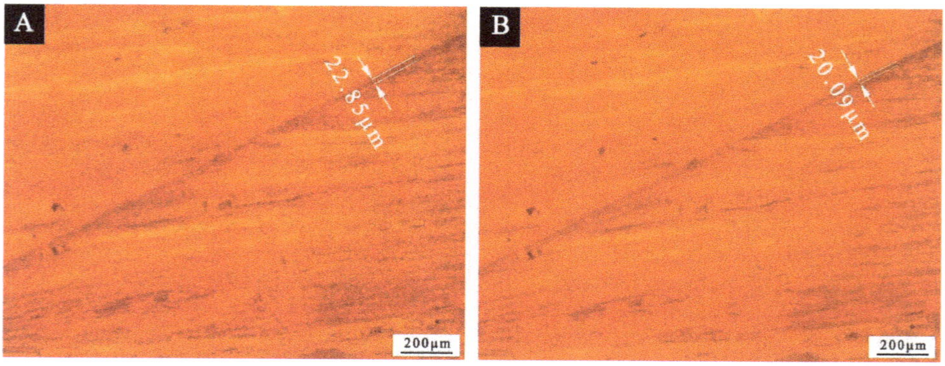

Figure 10. The OM diagram of paint coating without microcapsule: (**A**) 0 d, (**B**) 5 d.

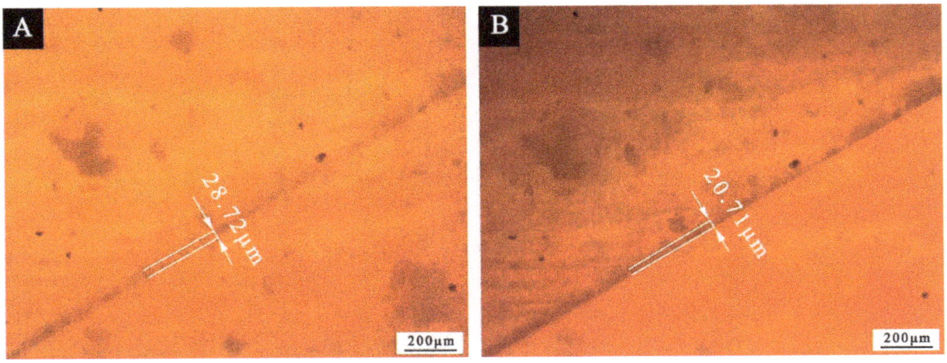

Figure 11. The OM diagram of paint coating with 10.0% microcapsule: (**A**) 0 d, (**B**) 5 d.

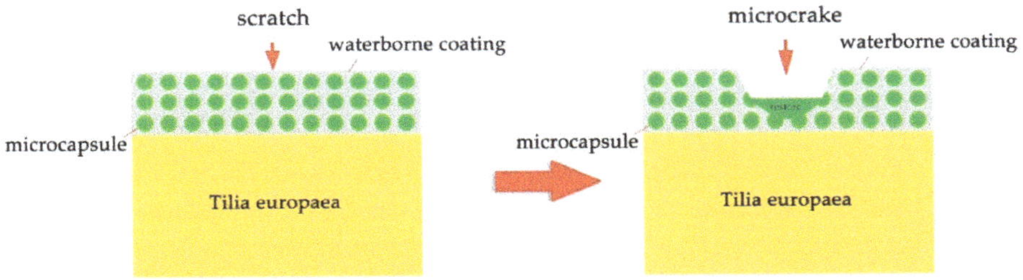

Figure 12. The self-repairing mechanism of waterborne coating microcapsules.

The performances of the waterborne coatings with the same content, the same wall material and different core materials reported in the literature were compared and analyzed [32–34]. In Table 10, the optimal mass ratio of core to shell with waterborne coating microcapsules was 0.67:1, that of waterborne acrylic acid microcapsules was 0.58:1, that of epoxy resin microcapsules was 0.83:1, and that of shellac microcapsules was 0.75:1. The results showed that the optimal core-wall ratio may be different for different kinds of core materials. From Table 10, it can be found that the optical and mechanical properties of waterborne coatings were significantly different after adding microcapsules with different core materials. Among them, the coating with waterborne coating microcapsules had the highest gloss, the coating with epoxy resin microcapsules had a higher elongation at break, and the coating with shellac microcapsule had the lowest hardness. It may be that epoxy resin is more flexible than waterborne coatings and shellac, so the elongation at break is higher. But matched with epoxy resin, the waterborne coating and shellac can be solidified at normal atmospheric temperature without heat, which is more practical for surface coating of wood furniture and wood products. However, the shellac is insoluble in water and soluble in organic solvents, which has certain difficulties in the preparation process. The coating with waterborne coating as the core material has better comprehensive properties.

4. Conclusions

Through orthogonal test of three factors and two levels, UF resin cladding waterborne coating microcapsules were prepared. The best reaction temperature is 70 °C and the best standing time is 5 d. The microcapsule with a mass ratio of core to shell of 0.67:1 has better comprehensive properties such as morphology, output, and clad ratio. Under the optimum preparation conditions, the clad ratio was 49.0%. Compared with the coating without microcapsules, the hardness of the coating with 10.0% mass fraction microcapsules increased to 3H, the impact strength increased to 13 kg·cm, and the elongation at break increased to 16.2%. Moreover, the microcapsules have a certain self-healing effect on micro cracks in wood surface coating. At the same time, compared with the coating with other different core materials, the coating prepared by adding microcapsules with waterborne coating as core material has better comprehensive properties. This may lay a technical foundation for the industrial application of waterborne coating microcapsules on wood surface. The future research direction is to prepare microcapsules with higher coverage rate and better repair effect for waterborne coatings on wood surface, at the same time, the wall materials will be environmental protection materials without formaldehyde emission.

Author Contributions: Conceptualization, methodology, validation, resources, data management, supervision, X.Y.; Formal analysis, investigation, writing-review and editing, Y.T.; Investigation and writing-review, X.Q. All authors have read and agreed to the published version of the manuscript.

Funding: This project was partly supported by the Natural Science Foundation of Jiangsu Province (BK20201386) and Youth Science and Technology Innovation Fund of Nanjing Forestry University (CX2016018).

Institutional Review Board Statement: Not applicable.

Informed Consent Statement: Not applicable.

Data Availability Statement: Not applicable.

Conflicts of Interest: The authors declare no conflict of interest.

References

1. Panek, M.; Oberhofnerova, E.; Zeidler, A.; Sedivka, P. Efficacy of Hydrophobic Coatings in Protecting Oak Wood Surfaces during Accelerated Weathering. *Coatings* **2017**, *7*, 172. [CrossRef]
2. Rowell, R.; Bongers, F. Role of Moisture in the Failure of Coatings on Wood. *Coatings* **2017**, *7*, 219. [CrossRef]
3. Panek, M.; Dvorak, O.; Oberhofnerova, E.; Simunkova, K.; Zeidler, A. Effectiveness of Two Different Hydrophobic Topcoats for Increasing of Durability of Exterior Coating Systems on Oak Wood. *Coatings* **2019**, *9*, 280. [CrossRef]
4. Liu, Q.; Gao, D.; Xu, W. Influence of the Bottom Color Modification and Material Color Modification Process on the Performance of Modified Poplar. *Coatings* **2021**, *11*, 660. [CrossRef]
5. Lee Hia, I.; Chan, E.-S.; Chai, S.-P.; Pasbakhsh, P. A Novel Repeated Self-healing Epoxy Composite with Alginate Multicore Microcapsules. *J. Mater. Chem. A* **2018**, *6*, 8470–8478. [CrossRef]
6. Gao, L.; He, J.; Hu, J.; Wang, C. Photoresponsive Self-Healing Polymer Composite with Photoabsorbing Hybrid Microcapsules. *ACS Appl. Mater. Inter.* **2015**, *7*, 25546–25552. [CrossRef] [PubMed]
7. Xu, W.; Fang, X.; Han, J.; Wu, Z.; Zhang, J. Effect of Coating Thickness on Sound Absorption Property of Four Wood Species Commonly Used for Piano Soundboards. *Wood Fiber Sci.* **2020**, *52*, 28–43. [CrossRef]
8. Garcia-Jimeno, S.; Cano-Sarabia, M.; Mejias, N.; Navarro, V.; Frances, A.B.; Maspoch, D. Healing Damaged Coatings Using Friction-sensitive Hybrid Microcapsules. *J. Mater. Chem. A* **2015**, *3*, 17966–17970. [CrossRef]
9. Wan, H.; Song, D.; Li, X.; Zhang, D.; Gao, J.; Du, C. A New Understanding of the Failure of Waterborne Acrylic Coatings. *RSC Adv.* **2017**, *7*, 38135–38148. [CrossRef]
10. Zhao, G.; Ding, C.; Pan, M.; Zhai, S. Fabrication of NCC-SiO_2 Hybrid Colloids and Its Application on Waterborne Poly(Acrylic Acid) Coatings. *Prog. Org. Coat.* **2018**, *122*, 88–95. [CrossRef]
11. Sun, Y.; Wang, C.; Wu, Y.; Zuo, J.; Zhan, X. Effect of Nano-boron Carbide on the Properties of Waterborne Polyurethane Wood Coatings. *J. For. Eng.* **2020**, *5*, 181–185.
12. Cai, Z.; Zhu, H.; Wang, P.; Wu, C.; Gao, W.; Mu, J.; Wei, S. Performance Optimization of UV Curable Waterborne Polyurethane Acrylate Wood Coatings Modified by Castor Oil. *J. For. Eng.* **2020**, *5*, 89–95.
13. Aguirreurreta, Z.; de la Cal, J.C.; Leiza, J.R. Preparation of High Solids Content Waterborne Acrylic Coatings Using Polymerizable Surfactants to Improve Water Sensitivity. *Prog. Org. Coat.* **2017**, *112*, 200–209. [CrossRef]
14. Jiang, L.; Shen, J.; Zhao, Z.; Dong, H.; Li, Y. Study on Film Properties and VOCs of Nano-TiO_2 and ZnO Modified Waterborne Paints. *J. For. Eng.* **2019**, *4*, 148–155.
15. Lin, X.; Su, J.; Feng, B.; Guo, S.; Chen, Y.; Liu, P.; Wei, S. Modification of Waterborne Acrylate Coatings Using Biomass Silicon. *J. For. Eng.* **2019**, *4*, 148–154.
16. Zhu, X.; Liu, Y.; Li, Z.; Wang, W. Thermochromic Microcapsules with Highly Transparent Shells Obtained through In-situ Polymerization of Urea Formaldehyde around Thermochromic Cores for Smart Wood Coatings. *Sci. Rep.* **2018**, *8*, 4015. [CrossRef]
17. Hu, L. Development of Photochromic Wood Material by Microcapsules. *BioResources* **2016**, *11*, 9547–9559. [CrossRef]
18. Jeong, S.-G.; Jeon, J.; Seo, J.; Lee, J.-H.; Kim, S. Performance Evaluation of the Microencapsulated PCM for Wood-based Flooring Application. *Energ. Convers. Manag.* **2012**, *64*, 516–521. [CrossRef]
19. Pinkl, S.; van Herwijnen, H.W.G.; Veigel, S.; Gindl-Altmutter, W.; Riegler, M. Urea-formaldehyde Microspheres as A Potential Additive to Wood Adhesive. *J. Wood Sci.* **2018**, *64*, 390–397. [CrossRef]

20. He, S. Microwave Treatment for Enhancing the Liquid Permeability of Chinese Fir. *BioResources* **2014**, *9*, 1924–1938. [CrossRef]
21. Wu, S.; Tao, X.; Xu, W. Thermal Conductivity of Poplar Wood Veneer Impregnated with Graphene/Polyvinyl Alcohol. *Forests* **2021**, *12*, 777. [CrossRef]
22. Ma, Y.; Zhang, Y.; Liu, X.; Gu, J. Synthesis of Isocyanate Microcapsules as Functional Crosslinking Agent for Wood Adhesive. *J. Adhes.* **2019**, *97*, 38–52. [CrossRef]
23. Liu, Y. Self-assembly of Poly (Styrene-methyl Methacrylate-acrylic Acid) (P(St-MMA-AA)) Colloidal Microspheres on Wood Surface by Thermal-assisted Gravity Deposition. *Wood Sci. Technol.* **2021**, *55*, 403–417. [CrossRef]
24. Liu, Y.; Hu, J. Investigation of Polystyrene-based Microspheres from Different Copolymers and Their Structural Color Coatings on Wood Surface. *Coatings* **2021**, *11*, 14. [CrossRef]
25. Yan, X.; Qian, X.; Wu, Z. Self-repairing Technology of Microencapsulate and its Applications in Coatings. *J. For. Eng.* **2019**, *4*, 20–28.
26. Siva, T.; Sathiyanarayanan, S. Self Healing Coatings Containing Dual Active Agent Loaded Urea Formaldehyde (UF) Microcapsules. *Prog. Org. Coat.* **2015**, *82*, 57–67. [CrossRef]
27. Lang, S.; Zhou, Q. Synthesis and Characterization of Poly(urea-formaldehyde) Microcapsules Containing Linseed Oil for Self-healing Coating Development. *Prog. Org. Coat.* **2017**, *105*, 99–110. [CrossRef]
28. Yang, H.-I.; Kim, D.-M.; Yu, H.-C.; Chung, C.-M. Microcapsule-Type Organogel-Based Self-Healing System Having Secondary Damage Preventing Capability. *ACS Appl. Mater. Inter.* **2016**, *8*, 11070–11075. [CrossRef]
29. Liu, Y.; Hu, J.; Wu, Z. Fabrication of Coatings with Structural Color on A Wood Surface. *Coatings* **2020**, *10*, 32. [CrossRef]
30. Jiao, C.; Shao, Q.; Wu, M.; Zheng, B.; Guo, Z.; Yi, J.; Zhang, J.; Lin, J.; Wu, S.; Dong, M.; et al. 2-(3,4-Epoxy) Ethyltriethoxysilane-modified Waterborne Acrylic Resin: Preparation and Property Analysis. *Polymer* **2020**, *190*, 122196. [CrossRef]
31. Liu, M.; Mao, X.; Zhu, H.; Lin, A.; Wang, D. Water and Corrosion Resistance of Epoxy–acrylic–amine Waterborne Coatings: Effects of Resin Molecular Weight, Polar Group and Hydrophobic Segment. *Corros. Sci.* **2013**, *75*, 106–113. [CrossRef]
32. Yan, X.; Tao, Y.; Qian, X. Preparation and Optimization of Waterborne Acrylic Core Microcapsules for Waterborne Wood Coatings and Comparison with Epoxy Resin Core. *Polymers* **2020**, *12*, 2366. [CrossRef] [PubMed]
33. Yan, X.; Qian, X.; Chang, Y. Preparation and Characterization of Urea Formaldehyde @ Epoxy Resin Microcapsule on Waterborne Wood Coatings. *Coatings* **2019**, *9*, 475. [CrossRef]
34. Yan, X.; Wang, L. Preparation of Shellac Resin Microcapsules Coated with Urea Formaldehyde Resin and Properties of Waterborne Paint Films for Tilia amurensis Rupr. *Membranes* **2020**, *10*, 278. [CrossRef] [PubMed]
35. Yan, X.; Chang, Y.; Qian, X. Preparation and Self-Repairing Properties of Urea Formaldehyde-Coated Epoxy Resin Microcapsules. *Int. J. Polym. Sci.* **2019**, *2019*, 1–11. [CrossRef]
36. Liu, Q.; Gao, D.; Xu, W. Effect of Sanding Processes on the Surface Properties of Modified Poplar Coated by Primer Compared with Mahogany. *Coatings* **2020**, *10*, 856. [CrossRef]

polymers

MDPI

Article

Preparations of Poly(lactic acid) Dispersions in Water for Coating Applications

Giada Belletti [1,2], Sara Buoso [2], Lucia Ricci [3], Alejandro Guillem-Ortiz [4], Alejandro Aragón-Gutiérrez [4], Olga Bortolini [1] and Monica Bertoldo [1,2,*]

1 Department of Chemical, Pharmaceutical and Agricultural Sciences, University of Ferrara, Via L. Borsari 46, 44121 Ferrara, Italy; giada.belletti@unife.it (G.B.); olga.bortolini@unife.it (O.B.)
2 Institute of Organic Synthesis and Photoreactivity, National Research Council, Via P. Gobetti 101, 40129 Bologna, Italy; sara.buoso@isof.cnr.it
3 Institute for Chemical and Physical Processes, National Research Council, Via G. Moruzzi 1, 54124 Pisa, Italy; lucia.ricci@pi.ipcf.cnr.it
4 Instituto Tecnológico del Embalaje, Transporte y Logística, ITENE, Calle de Albert Einstein 1, 46980 Paterna, Spain; alejandro.guillem@itene.com (A.G.-O.); alejandro.aragon@itene.com (A.A.-G.)
* Correspondence: monica.bertoldo@unife.it

Abstract: A green, effective methodology for the preparation of water-based dispersions of poly(lactic acid) (PLA) for coating purposes is herein presented. The procedure consists of two steps: in the first one, an oil-in-water emulsion is obtained by mixing a solution of PLA in ethyl acetate with a water phase containing surfactant and stabilizer. Different homogenization methods as well as oil/water phase ratio, surfactant and stabilizer combinations were screened. In the second step, the quantitative evaporation of the organic provides water dispersions of PLA that are stable, at least, over several weeks at room temperature or at 4 °C. Particle size was in the 200–500 nm range, depending on the preparation conditions, as confirmed by scanning electron microscope (SEM) analysis. PLA was found not to suffer significant molecular weight degradation by gel permeation chromatography (GPC) analysis. Furthermore, two selected formulations with glass transition temperature (Tg) of 51 °C and 34 °C were tested for the preparation of PLA films by drying in PTFE capsules. In both cases, continuous films that are homogeneous by Fourier-transform infrared spectroscopy (FT-IR) and SEM observation were obtained only when drying was performed above 60 °C. The formulation with lower Tg results in films which are more flexible and transparent.

Keywords: poly(lactic acid); water emulsions; water dispersions; film formation

Citation: Belletti, G.; Buoso, S.; Ricci, L.; Guillem-Ortiz, A.; Aragón-Gutiérrez, A.; Bortolini, O.; Bertoldo, M. Preparations of Poly(lactic acid) Dispersions in Water for Coating Applications. *Polymers* **2021**, *13*, 2767. https://doi.org/10.3390/polym13162767

Academic Editor: George Z. Kyzas

Received: 30 June 2021
Accepted: 12 August 2021
Published: 18 August 2021

1. Introduction

In recent years, in an effort to produce more eco-friendly products, renewable and biodegradable polymers have been investigated as emerging coating materials. One of the most promising polymers for such a purpose is poly(lactic acid) (PLA) due to its biodegradability, biocompatibility and production on an industrial scale at relatively low cost [1–3]. PLA is already used for a wide range of applications, particularly in the packaging sector [4], for instance for making food and beverage containers, cups, overwrap, blister packages, as well as coating on paper and board [5–7]. The most frequently used conversion method for the preparation of PLA items is melt processing, such as cast extrusion, extrusion coating, lamination and blown extrusion [8,9]. An alternative technology for the preparation of coatings that does not involve treatment at high temperature is solvent casting [10,11]. The method requires the solubilization of the polymer in a suitable solvent, followed by casting of the solution onto the substrate. Despite the possibility to modulate the film thickness by varying the concentration of the polymer solution, the release of toxic and harmful organic solvents during the process gave rise to health and environmental issues. For this reason, the latter method is not typically used to produce biodegradable films. Indeed, in the last decades, many efforts have been devoted to the replacement of solvent-borne polymer coatings with their waterborne counterparts, even for those applications where they were traditionally used such as adhesives and inks [12,13]. Water, indeed, is considered as a cheap, safe, non-toxic and environmentally benign solvent [14]; however, polymers are rarely soluble in water. A typical method to use water as a vehicle, in spite of the absence of solubility, is the preparation of polymer latexes consisting of polyacrylate, polymethacrylate and more recently, polyurethane as well [15–22]. Water dispersions are gaining land in many application fields such as paints [23], adhesives [24] and inks [25,26], particularly for the food packaging sector [27,28], where the restrictions on the possible substances used forced the producers to substitute old technologies with safer ones.

When water dispersions (or latexes) are used to prepare coatings, the preparation is cast onto the chosen substrate and, as water evaporates, the polymer particles merge and form a homogenous film on the coated surface [29]. The ability of a polymer latex to form a continuous film depends on the minimum film formation temperature (MFFT) of the polymer. When the casting takes place above the MFFT, the particles undergo deformation and cohesion, thus creating a homogeneous film. On the other hand, when the film is cast below its MFFT, a brittle discontinuous film or compact powder may be obtained [30]. Typically, the MFFT tends to be close to the glass transition temperature (Tg) of the polymer or above it [31]. The most commonly accepted mechanism of film formation from waterborne dispersions of preformed polymers is composed of three stages: drying, deformation and coalescence [32,33]. During drying, the water evaporates from the dispersion surface at a constant rate until the polymer has reached 60–70% of volume fraction. As the water evaporation rate slows down, deformation occurs and the particles come into irreversible contact. In the final step, polymer chain interdiffusion takes place and the particles coalesce into a continuous film.

The most common methods to produce water-based coatings are emulsion polymerization [34] and emulsification of preformed polymers. Polymer emulsions of acrylate and methacrylate have been produced for many years with emulsion polymerization [34–38]. More recently, water dispersion of low molecular weight resins that are converted into high weight mechanically performing coatings by UV-curing have also entered into the market [39,40]. Unluckily, emulsion polymerization is applicable only to free radical polymerizable monomers and lactide and lactic acid are not included among them. Indeed, in these cases the production of water-based preparations is achieved by dispersing the preformed polymer [41]. For instance, in the acetone process for the preparation of polyurethane-polyurea water dispersions, the hydrophilic isocyanate prepolymer is firstly chain extended in acetone to give the desired molecular weight. Subsequently, by the addition of water and removal of the solvent, a purely waterborne dispersion is formed [42,43].

To the best of our knowledge the preparation of aqueous emulsions and dispersions of PLA has been reported only for applications in the biomedical field, especially for the obtainment of PLA nanoparticles for drug delivery [44,45]. Some of the methods used in these cases involve the use of non-harmful solvents, such as acetone and ethyl acetate. Very diluted dispersions of PLA particles have been obtained by nanoprecipitation [46–49], salting-out [41,50–52] and emulsification-diffusion methods [46,53,54]. Nevertheless, these works aim at the isolation through centrifugation or lyophilization of the PLA nanoparticles. As a consequence, the dispersion stability over time is usually not investigated and besides the final PLA content is usually lower than 5 wt.%. On the other hand, with the use of chlorinated solvents by emulsification-evaporation methods, PLA dispersions with higher polymer contents are accessible (maximum 20 wt.%) [27,55–57]. Generally, the simplest emulsion-based procedures require the solubilization of PLA in a water-immiscible organic solvent, followed by subsequent dispersion of this organic phase in fine droplets into the aqueous medium. After that, the organic solvent evaporates leading to the formation of PLA nanoparticles dispersed in water [44]. This technique was also exploited to prepare blends of PLA with hydrophilic polymers such as chitosan [58,59], lignin [60] and nanocellulose [61,62]. However, in the latter cases, the studies focus on the preparation of thermoplastic composites and consequently, the stability of the dispersion was not an issue and thus it was not reported. Besides, the toxicity associated with the chlorinated solvents used in these processes limits the biocompatibility and the use of the dispersions thus prepared for food contact applications. Additionally, the preparation of waterborne PLA dispersions for coating applications has been the subject of some recent patents [63,64]. In these cases, preparations with high dry content (25–70 wt.%) are afforded in the absence of organic solvent by continuously and simultaneously introducing melted PLA and aqueous phase into the blender of an emulsifying unit. Nevertheless, with these techniques the polymer is always processed at high temperatures and usually, high amounts of viscosity reducing agents, plasticizers or additives are needed. The PLA particles thus obtained are in the micrometric size range.

Aim of the present work was the design and optimization of a process to obtain stable dispersions of PLA in water suitable for coating film formation. The process had to be accomplished without chlorinated solvents to allow the use of the produced coatings in food packaging application. To this aim, an amorphous PLA soluble in non-chlorinated solvents and a solvent approved for food contact application, such as ethyl acetate, were selected. Different homogenization methods were tested for the mixing of the organic phase, in which PLA was dissolved, and the water-phase containing emulsifier and stabilizer. With the purpose of ensuring the stability of the preparation during storage and during evaporation of the organic solvent, the use of several emulsifiers and different polysaccharides as stabilizers were examined. In fact, to prepare kinetically stable emulsions, an emulsifier is needed to protect the newly formed droplets against the different destabilization mechanisms. In particular, the emulsifier forms a protective interfacial layer when it adsorbs on the surface of the droplets, preventing the latter from merging together [65]. Emulsifiers can be produced using either petrochemical or bio-based feedstocks and they are commonly classified according to the polarity of their head group: anionic, cationic, non-ionic and amphoteric [66]. In recent years, due to the continuous consumption and depletion of fossil fuels, renewable substrates have been of increasing interest for the production of bio-based surfactants [67]. On the other hand, non-surface-active polysaccharides, like gelatinized starch, may contribute to interfacial stabilization of an oil-in-water emulsion by interaction with the surfactant layer already located at the interface [65,68]. Additionally, polysaccharides are able to enhance the stability of emulsions by increasing the viscosity of the aqueous phase [69]. In this work, various chemicals and polysaccharides approved for food contact application were chosen as promising surfactants to be tested. Furthermore, several emulsifiers were selected on the basis of the relative strength of their hydrophilic and lipophilic moieties (known as hydrophilic–lipophilic balance or HLB) [70]. Surfactants with both high and low HLB values were examined. Regarding the choice of the stabilizers, the attention was focused on bio-based polysaccharides, which are largely employed in the food industry such as starch and xanthan gum (XG) [71,72] or promising emulsion stabilizers such as microfibrillated cellulose (MFC) [73]. Additionally, the film forming ability of two selected PLA dispersion formulations was studied at different temperatures in the 25–110 °C range. The morphology of the obtained layers was investigated through scanning electron microscopy (SEM) to identify the MFFT of the innovative biodegradable dispersions in water.

2. Materials and Methods

2.1. Materials

PLA Ingeo Biopolymer 4060D (PLA) was supplied by NatureWorks LLC (Blair, NE, USA.). Sodium dodecyl sulfate (SDS), ethyl acetate, sodium alginate (SA), sodium carboxymethylcellulose (NaCMC), Tween 80, Tween 20, stearoyl-2-lactylate (SSL) and Span80 were purchased from Sigma-Aldrich (Milan, Italy). Starch C*Icoat 07,525 (starch) and xanthan gum (XG) were generously supplied by Cargill (Krefeld, Germany,). A solution of starch at 3% in ultrapure water was prepared by heating under reflux 6 g in 200 mL for nearly 1 h. Metolad 368 (Met. 368), Metolad 388 (Met. 388), Leukonol LB-2 (Leu.) and Tafigel AP 15 (Taf.) were kindly provided by Münzing Chemie GmbH (Abstatt, Germany). Synperonic PE/F68 (SYN) was generously supplied by Croda Chemicals (Mortara (PV), Italy). Microfibrillated cellulose MFC Exilva F 01-L (MFC) was kindly supplied by Borregaard (Sarpsborg, Norway).

2.2. Preparation of PLA Dispersions with Sodium Dodecyl Sulfate as Surfactant and Starch as Stabilizer

6.47 g of starch solution at 3% conc. was diluted with 28.5 mL of ultrapure water (starch conc. 0.5 wt.%) and 0.28 g of SDS were added (SDS conc. 0.8 wt.%). In the meantime, 5 or 7 g of PLA was solubilized in 45 or 87.5 mL of ethyl acetate (EtOAc) (11 or 8 wt./vol.%, respectively) under vigorous stirring for 4 h at room temperature. After that, proper amounts of the oil and water phases were mixed, in an ice bath, with the use of a homogenizer to obtain the emulsions. Different tests were carried out by changing the volume ratio between the two phases as well as the total phase volume (entries 1–13, Table 1). For homogenization, ULTRA-TURRAX (UT) or two different ultrasound probes (UVC and UH) were tested. Homogenization conditions were as follows: (a) UH for 30 s at amplitude 50% + 1 min 30 s at amplitude 80%; (b) UVC for 30 s at amplitude 50% + 2 min of overhead stirrer at 200 rpm + (30 s at amplitude 90% + 2 min of overhead stirrer at 200 rpm) three times; (c) UT for 1 min 30 s at 1100 rpm. Afterwards, the resulting emulsions were stirred for 20 h at room temperature and 200 rpm under the aspiration of a laboratory hood (suction speed: 0.5 m/s).

A similar procedure was adopted in tests of entry 11 and 12 in Table 1 except that the amount of the starch solution used was 0.38 g (starch conc. 1.1 wt.%) in test 11 and the amount of SDS employed in test 12 was 0.10 g (SDS conc. 0.3 wt.%).

Table 1. Experimental conditions adopted to prepare ethyl acetate-(EtOAc) in-water PLA dispersions with SDS as surfactant and starch as stabilizer.

Entry	Organic Phase		$EtOAc/H_2O$	SDS Conc. (wt.%)	Vol.Tot. (mL)	PLA/Water [2] (wt.%)	Dispersion Features		
	PLA (wt./vol.%) [1]	Vol. (mL)	Vol. Ratio				Homogenizer [3]	Stability [4]	Particles Size [5] (nm)
1	11	45.0	1.3	0.8	80	14	UT	coa.	-
2	11	45.0	1.3	0.8	80	14	UH	yes	214 ± 2
3	11	45.0	1.3	0.8	80	14	UVC	yes	203 ± 7
4	11	62.5	1.8	0.8	97	20	UVC	cl.	225 ± 3
5	8	62.5	1.8	0.8	97	14	UVC	yes	178 ± 2
6	8	20.8	1.8	0.8	32	14	UVC	yes	167 ± 1
7	8	87.5	2.5	0.8	122	20	UVC [6]	cl.	202 ± 2
8	8	87.5	2.5	0.8	122	20	UH [6]	yes	130 ± 10
9	8	50.0	2.5	0.8	70	20	UVC	coa.	-
10	8	50.0	2.5	0.8	70	20	UH [6]	yes	138 ± 23
11	8	50.0	5.0	0.8	60	15	UH	coa.	-
12	8	50.0	2.1	0.3	74	17	UH	yes	158 ± 12
13	8	20.8	1.8	0.5	58	15	UVC	yes	183 ± 2

[1] Ratio between the weight of poly(lactic acid) and the volume of EtOAc. [2] Ratio between the weight of poly(lactic acid) and the volume of water. [3] UH = ultrasonication with Hielscher, UVC = ultrasonication with Vibracell, UT = Ultraturrax. [4] Assessed through visual criteria and DLS analysis both after the preparation and after 1 week. Stability as follows: coa. = coagulation, cl. = clot formation. [5] Determined through DLS analysis. [6] 1.1 wt.% of starch in water was employed.

2.3. Preparation of PLA Dispersions with SDS and Different Polysaccharides as Stabilizers

Different amounts of starch, MFC or XG and 0.28 g of SDS were solubilized in 35 mL of ultrapure water (Table 2). In the case of MFC, the polysaccharide suspension was treated with UT at 10,000 rpm for 4 min before addition of SDS. In the meantime, 5 g of PLA was solubilized in 45 or 62.5 mL of EtOAc (11 or 8 wt./vol.%, respectively) under vigorous stirring for 4 h at room temperature. After that, the oil and water phases were mixed together, in an ice bath, with the aid of the UVC homogenizer for 30 s at amplitude 50% + 2 min of overhead stirrer at 200 rpm + (30 s at amplitude 90% + 2 min of overhead stirrer at 200 rpm) three times. Afterwards, the resulting emulsion was stirred for 20 h at room temperature at 200 rpm under the aspiration of a laboratory hood (suction speed: 0.5 m/s).

Table 2. Screening of different stabilizers for the preparation of PLA dispersions in the presence of SDS as surfactant.

	Aqueous Phase		Organic Phase		$EtOAc/H_2O$	Dispersion Feature
Entry	**Stabilizer**	**wt.%**	**PLA (wt./vol.%)** [1]	**Vol. (mL)**	**Vol. Ratio**	**Stability** [2]
1	-	-	11	45.0	1.3	no
2	-	-	8	62.5	1.8	no
3	MFC	0.500	11	45.0	1.3	sed.
4	XG	0.090	11	15.0	1.3	no
5	XG	0.009	11	15.0	1.3	yes
6	XG	0.009	11	45.0	1.3	coa.

[1] Ratio between the weight of poly(lactic acid) and the volume of EtOAc. [2] Assessed through visual criteria and DLS analysis both after the preparation and after 1 week. Stability as follows: coa. = coagulation, sed. = sedimentation.

2.4. Preparation of PLA Dispersions with Different Surfactants Than Sodium Dodecyl Sulfate

The proper amount of emulsifier was solubilized into 35 mL of ultrapure water. In the case of the presence of starch as stabilizer in the formulation, the 35 mL included 6.47 g of starch solution at 3% conc. (starch conc. 0.5 wt.%). The type of surfactant employed and its concentration in the water phase are listed in Table 3.

Table 3. Screening of different surfactants for the preparation of PLA dispersions.

Entry	Surfactant	Description	Aqueous Phase			Dispersion Stability [1]
			Surfactant Conc. (wt.%)	**HLB**	**Starch Conc. (wt.%)**	
1	SDS	Anionic surfactant	0.8	40	0	unstable
2 [2]	SDS Span 80	Anionic surfactant Non-ionic surfactant	2.5 0.5	34	0	stable
3	SYN	Block copolymer PEO/PPO	4.8	29	0	unstable
4	SYN		2.4	29	0	stable
5	SYN		2.0	29	0	unstable
6	SYN		1.6	29	0	unstable
7	SYN		0.8	29	0	unstable
8	SYN Span 80	Block copolymer PEO/PPO Non-ionic surfactant	0.8 0.5	19	0	unstable
9	Tween 80	Non-ionic surfactant	0.5	15	0	partial
10	SYN	Block copolymer PEO/PPO	2.4	29	0.5	stable
11	SYN		0.8	29	0.5	unstable
12	Met. 388	Non-ionic compound	0.8	19	0.5	inverted
13	Tween 20	Non-ionic surfactant	0.5	17	0.5	partial
14	Tween 80	Non-ionic surfactant	1.0	15	0.5	partial
15	SA	Anionic polysaccharide	0.8	11	0.5	inverted
16	NaCMC	Anionic cellulose derivative	0.8	11	0.5	inverted
17	SSL	Anionic surfactant	0.8	9	0.5	inverted
18	Met. 368	Ester based	0.8	8	0.5	inverted

[1] Assessed through visual criteria and DLS analysis both after the preparation and after 1 week. Inverted = formation of water-in-oil emulsion instead of oil-in-water one. [2] Homogenization with UH, PLA/EtOAc ratio (wt./vol.%) = 8, $EtOAc/H_2O$ vol. ratio = 2.5, total volume = 70 mL.

Moreover, 5 g of PLA was dissolved in 45 mL of EtOAc under vigorous stirring for 4 h at room temperature. When Span 80 was employed as surfactant, the latter was solubilized in the organic phase together with PLA. After that, the oil and water phases were mixed together and homogenized while standing in an ice bath to obtain the emulsions. Homogenization was carried out with UH or UVC ultrasound homogenizer under conditions "a" or "b" mentioned in Section 2.2, respectively. Afterwards, the resulting emulsions were stirred overnight at 200 rpm under the aspiration of a laboratory hood (suction speed: 0.5 m/s).

2.5. Preparations of PLA Films from Dispersions

1.7–3.0 g of PLA dispersions prepared under the condition of entry 3, Table 1 (PLA_SDS_starch) or entry 4, Table 3 (PLA_SYN), were transferred into a PTFE capsule (diameter 3.7 cm) and allowed to dry for 2–3 h at the selected temperature until solidification occurred. The drying temperature was room temperature and then the latter was gradually increased by 10 °C starting from 40 up to 110 °C.

2.6. Instruments and Characterization Methods

To weigh the starting materials Entris Sartorius and Sartorius BP61S laboratory balances (Sartorius AG, Goettingen, Germany) were used. To prepare the water and the organic phase of the dispersions, an IKA magnetic stirrer RCT basic (IKA, Staufen, Germany) was employed. Ultrapure water was produced with Millipore Direct-Q 3UV (Merck KGaA, Burlington, MA, USA).

For homogenization of oil and water phases, Ultrasound Hielscher Up200St (UH) equipped with a 14 mm diameter probe, Vibra-Cell Ultrasonic Liquid Processors VCX750 (UVC) (Sonics, Newtown, MA, USA) provided with a 13 mm diameter probe and IKA ULTRA-TURRAX T 25 basic (UT) equipped with an S 25 N-18 G dispersing tool (IKA, Staufen, Germany) were used.

Emulsions and dispersions were stirred with an ArgoLab AM20-D (ArgoLab group, Landshut, Germany) overhead stirrer equipped with PTFE blade.

The diameter of the particles was determined by dynamic light scattering (DLS) analysis at 25 °C using a NanoBrook Omni particle size analyzer (Brookhaven Instruments Corporation, Holtsville, NY, USA) equipped with a 35-mW red diode laser (nominal 640 nm wavelength), BI-SCP cell in backscattering (173°) and software for data analysis. To give reliable results, dilute suspensions (0.15 vol./vol.%) were analyzed. For data elaboration, reflective indices of the dispersion medium and of the dispersed phase were assumed to be 1.330 and 1.596, respectively. Each measurement was repeated five times on the same sample and the reported data were the average over the five measurements. The standard deviation δ (Equation (1)) was assumed as data error, where N is the number of measurements, x_i the value of the i-th measurement and μ the arithmetic mean. All the DLS data were analyzed by a multimodal function. Particles were assumed to be uniform spheres.

$$\delta = \sum_{i=1}^{N} \frac{\sqrt{(x_i - \mu)^2}}{\sqrt{N}} \tag{1}$$

The dispersions' stability was assessed through visual criteria and DLS analysis both after the preparation and after 1 week. Dispersions were assumed stable if no phase separation was observed and the particle size variation was lower than the standard deviation on the particle size value.

The dry residuum of the preparations was determined after heating the sample at 200 °C for 30 min in an oven.

The residual amount of EtOAc present in the emulsion was quantitatively determined by gas chromatography/mass spectroscopy analysis (GC/MS). A triple quadrupole GC/MS instrument from Agilent Technologies (Agilent Technologies, Inc., Santa Clara, CA, USA) equipped with a 20 × 0.18 mm, 0.18 μm column was employed, using hydrogen at 0.8 mL/min as carrier gas with a 10:1 split ratio. For the analysis, 100 μL of the emulsion was added to a 20 mL chromatography vial and incubated at 85 °C for 30 min for head-space extraction. The GC oven was programmed for 40 °C for 3 min, followed by a stepped increase of 10 °C min^{-1} to 50 °C, where it was held for 5 min, and then the temperature was increased by 30 °C min^{-1} to 230 °C, where it was held for 3 min. The quantitative determination of EtOAc traces was conducted by using the selected ion monitoring (SIM) at the m/z values 88/73. Calibration was performed by injecting solutions of EtOAc in *N,N*-dimethylformamide at known concentrations into the GC/MS instrument.

Differential scanning calorimetric analyses were performed on a DSC 8000, PerkinElmer Inc. (Waltham, MA, USA) instrument equipped with IntraCooler II cooling device and Pyris software (Version 13.3, PerkinElmer Inc., Waltham, MA, USA) for instrument control, data acquisition and analysis. The instrument was calibrated for temperature and energy with high-purity indium and lead as standards. 3–10 mg of sample was analyzed in aluminum pans under dry nitrogen atmosphere (30 mL/min). Samples were at first heated up from 25 to 200 °C to erase the thermal history and to remove any trapped volatile substance such as residual solvents. Thus, samples were cooled down to −70 °C (cooling step), maintained at −70 °C for 5 min and finally heated up again to 200 °C (second heating step). Heating and cooling steps were all performed at 10 °C/min as the scanning rate.

Size-exclusion chromatography (SEC) analyses were performed with a Jasco (Jasco Europe srl, Cremella, Italy) instrument comprising a PU-2089 Plus quaternary pump and injector with a 20 mL loop, two in-series PLgel MIXED-D columns (Agilent Technologies Italia S.p.A., Cernusco sul Naviglio, Italy; linearity range 200 to 2,000,000 g/mol based on polystyrene equivalent) placed in a Jasco CO-2065 column oven set at 30 °C, a Jasco RI-2031 Plus refractive index detector, and a Jasco UV-2077 Plus multi-channel UV-Vis detector. The samples in the form of films or powders were dissolved in trichloromethane (HPLC grade Sigma-Aldrich, Milan, Italy) with the aid of sonication and filtered through a 0.2 mL pore size PTFE filter to remove the insoluble fraction before injection as 5 mg/mL solutions; elution was performed with trichloromethane at 1 mL/min flow rate. ChromNav Jasco software (Jasco Europe srl, Cremella, Italy) was used for data acquisition and analysis based on a calibration curve obtained by running a set of four monodisperse polystyrene standards (19,000, 50,000, 233,000, and 300,000 g/mol, respectively) and performing a 4th order fit.

Scanning electron microscopy analyses (SEM) were accomplished with a Zeiss EVO 40 microscopy (Carl Zeiss Microscopy Ltd., Cambridge, UK) equipped with a LaB_6 source. Samples in the form of powder or film were gold sputtered before observation. Film sections were obtained by fracture in liquid nitrogen.

A Perkin Elmer Lambda 650 UV-Vis spectrophotometer (Waltham, MA, USA) was used to record the transmittance of free-standing films. Measurements were accomplished in transmission mode with air as reference. Film thickness was measured with an electronic outside micrometer that is able to perform measurement between 0.001 and 25 mm.

FT-IR spectra were recorded on an Agilent Cary 630 FTIR spectrophotometer (Agilent Technologies, Inc., Santa Clara, CA, USA) with ZnSe ATR element. Background and sample spectra were collected by accumulating 32 scans.

3. Results and Discussion

3.1. Influence of the Preparation Conditions on the Stability of PLA Dispersions in Water

Several tests to prepare ethyl acetate/water (EtOAc/H_2O) emulsions of PLA with SDS as surfactant and starch as stabilizer were carried out (Table 1), by using Ultraturrax (UT) or Ultrasound processors (UH and UVC) as homogenizer. When Ultraturrax (UT) was used, the emulsion formed but it was unstable (entry 1, Table 1); during the subsequent evaporation of the organic solvent, the majority of the preparation coagulated. On the contrary, with the use of ultrasound processors, either Ultrasound Hielscher UP200St (UH) or Vibra-Cell Ultrasonic Liquid Processors VCX750–SONICS Materials (UVC), stable PLA emulsions in water were obtained (Figure 1a). As a matter of fact, by solvent evaporation, these were converted into stable dispersions with size of 200–215 nm. No significant effect of the specific ultrasound processor used on the particle size was noticed (Table 1). Dry matter content was between 13 and 20 wt.%, which is much higher than any other reported result for stable PLA dispersions in water produced with emulsification-evaporation methods and non-chlorinated solvents [29,55].

Inspired by the good results obtained with ultrasound homogenizers, we attempted to further increase the PLA content in water by using a larger volume of organic phase, namely, 62.5 mL instead of 45 mL, to prepare the emulsion with the UVC homogenizer (entry 4, Table 1). However, even if the emulsion formed, the process was not effective since a cloth made of PLA formed on the stirrer during the subsequent EtOAc evaporation. When the process was replicated with the same oil and water phase volumes, but with 8% concentration of PLA instead of 11% in the organic phase, once again a homogeneous emulsion was obtained.

(a)

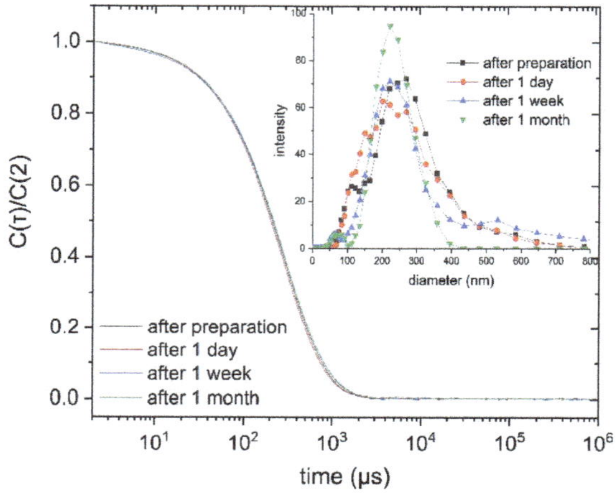

(b)

Figure 1. (**a**) Picture of the PLA dispersion of entry 5, Table 1; (**b**) comparison between DLS plots of the PLA dispersions prepared under the condition of entry 3, Table 1, after storing for increasing periods of time. Correlograms are normalized with respect to the intensity at time 2 μs. Insert refers to the particle size distribution as obtained by data fitting with a multimodal size distribution function.

In this case, the emulsion was stable enough to be converted into a stable dispersion by EtOAc evaporation. Particle size was smaller than the one obtained in the previous test with higher PLA concentration (entry 5, Table 1). Similar results were afforded when the total volume of the organic and water phases was reduced from 97 to 32 mL (entry 6, Table 1). The results suggest that the viscosity of the oil phase is the key parameter allowing the dispersion in the case of the higher oil/water phase ratio. Indeed, dispersion was achieved with PLA concentration of 11% and phase ratio of 1.3 or PLA concentration of 8% and phase ratio of 1.8. The decrease of the PLA concentration in EtOAc leads to a lower viscosity of the organic phase, thus reducing the energy input needed to disperse the organic phase into the aqueous one [74]; as a consequence, the treatment of a larger total volume becomes feasible. The further attempt to prepare a more concentrated PLA dispersion by increasing to 87.5 mL the volume of the 8% PLA solution (oil/water phase ratio of 2.5), once again gave an emulsion. However, this was not very stable and a clot formed on the stirrer during evaporation of the organic solvent, if the UVC homogenizer was used (entry 7, Table 1). On the contrary, with UH homogenizer, after solvent evaporation, the emulsion turned into a homogeneous dispersion with 20% dry matter content (entry 8, Table 1). Even when the preparation was performed in a smaller scale (with a lower total volume) but with the use of UVC homogenizer, a clot on the stirrer was formed during evaporation of the EtOAc (entry 9, Table 1). This result clearly indicates the higher efficiency of UH in transferring the energy and in treating larger both total volume and EtOAc/H_2O volume ratio than UVC does. Indeed, UH sonication was effective in treating EtOAc/H_2O volume ratio up to 2.5 (entries 2, 8, 10 and 11, Table 1), whilst the ratio was only up to 1.8 for preparations performed with UVC (entries 3, 4, 5, 6, 7 and 9, Table 1). In any case, the maximum EtOAc/H_2O volume ratio that could be treated is 2.5. As a matter of fact, no stable preparation could be obtained with a ratio of 5, even with the UH homogenizer, a relatively low total volume (60 mL) and a high concentration of stabilizer, 1.1% instead of 0.5% (entry 11, Table 1).

When the ratio between phases was lower than 2.5 for treatment with UH and not higher than 1.8 for UVC use, stable emulsions and then dispersions were obtained, even with SDS concentration lower than 0.8%. Indeed, good results were obtained with SDS 0.3% (entry 12, Table 1) plus 0.5% of starch.

3.2. *Influence of Ethyl Acetate Residua on the Emulsion Stability*

The homogeneous emulsions obtained in the experiments described in Table 1, if left standing without stirring, coagulated in a reasonably short time. However, it was clear that they turned into stable dispersions after complete evaporation of the organic solvent (Figure 1a). Dispersions can even be stored at room temperature or in the fridge at +4 °C for days and months without sedimentation and significant change of the particles' size and of their distribution (Figure 1b).

In order to better understand the emulsion preparation process, different samples were singled out at distinct times during the evaporation of the EtOAc and they were analyzed through DLS and head-space GC. Furthermore, the evolution of the dry matter content, while EtOAc evaporates, was followed. The experiment was carried out under the condition of entry 3 in Table 1. The evolution of the dry matter content shows an almost linear increase over time (Figure 2a). On the contrary, the EtOAc residue initially decreases slowly, and then almost exponentially. The difference among the two kinetics clearly indicates that during the evaporation stage not only the organic solvent but also a portion of water is removed from the preparation. This is due to the minimum temperature of the azeotrope between water and EtOAc.

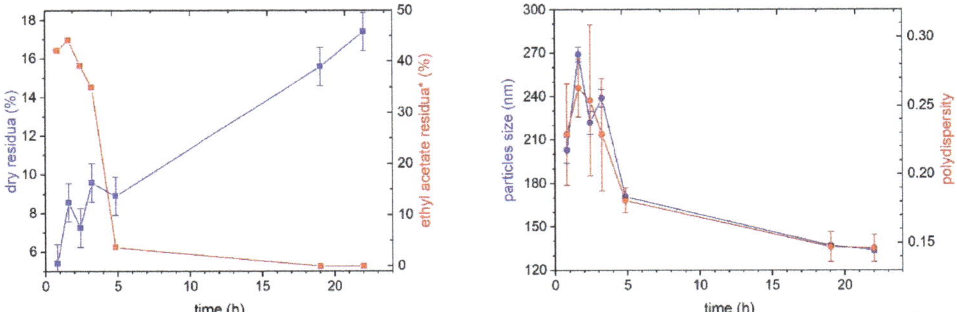

Figure 2. (**a**) Kinetics of ethyl acetate residue (red line) and dry residue (blue line) evolution; (**b**) kinetics of particles size (blue line) and polydispersity (red line) evolution over time. Data are related to test 3 in Table 1. * The highest data of ethyl acetate residua expressed in ppm are only indicative.

In the initial period (0–3 h) when the EtOAc residuum is very high (~40%) but also later on, when it is lower, e.g., 37% at 5 h, the preparation is unstable and coagulates by standing. On the contrary, when the EtOAc residuum reaches the value of 80 ppm, after 19 h, the preparation is almost stable; the DLS correlogram exhibits just a slight decrease of the intensity during the analysis time and the emulsion was visibly stable. Besides, when the EtOAc residuum is only 33 ppm (after 22 h), the dispersion is very stable, even after a month of storing in the fridge at 4 °C (Figure 1). Overall, the data collected show that, in order to ensure the stability of the dispersion, the organic solvent must be completely removed (residue of EtOAc lower than 40 ppm).

During EtOAc evaporation, particle size and polydispersity evolve, both passing through a maximum after 90–100 min and then decreasing to reach the minimum after 1 day (Figure 2b). The initial increase of size can be due to coalescence between distinct particles. Indeed, after preparation, the dispersed phase is made of drops of PLA solution in EtOAc. As the solvent evaporates, the particle volume decreases and then also its size. Furthermore, due to the evaporation of EtOAc, the concentration of PLA in the particle increases and therefore, the viscosity of the oil phase increases also. As a result, the rate of coalescence between distinct particles decreases and becomes irrelevant once particles are solid.

3.3. *Effect of Stabilizer on Dispersions Stability*

With the purpose of studying the role of starch on the stability of the emulsion/dispersion, different trials to prepare the emulsion with only SDS as surfactant or with other polysaccharides than starch were performed (Table 2). Two different tests were carried out with only SDS as surfactant. The water phase amount, composition and the total amount of PLA were the same. The only difference between the two tests was the concentration of the oil phase, which was 11% and 8%, (entries 1 and 2, Table 2, respectively), and as consequence the oil phase amount and the oil/water phase ratio also changed. In both cases, PLA dispersions unstable by DLS analysis were afforded. The comparison with the high stability of the dispersions prepared under comparable conditions with the additional presence of starch (entry 3 and 5 in Table 1, respectively) clearly indicates that the presence of such agent is of utmost importance for the stabilization of the resulting formulations. Usually, starch can act as an emulsifier only if hydrophobic modifications are introduced on its chain in order to afford substantial surface activity at the oil-water interface [75]. Gelatinized starch, as a non-surface-active polysaccharide, has been reported to act as stabilizer for the emulsion by forming a secondary steric stabilization layer through interaction with the pre-adsorber emulsifier [65,68].

After the result with starch, we decided to explore if similar results could be obtained with other polysaccharides. To this aim, microfibrillated cellulose and xanthan gum were tested. Microfibrillated cellulose (MFC) in water has been reported to form a three-dimensional network capable of incorporating dispersed oil droplets, preventing their coagulation and sedimentation [76]. On the other hand, xanthan gum (XG) is widely used to increase the viscosity of the aqueous phase, hence preventing creaming of oil droplets [77]. However, in the case of the PLA, the attempt to employ MFC as stabilizer in a concentration of 0.5 wt.% in water, in the presence of SDS, gave a PLA dispersion that sedimented over time (entry 3, Table 2). On the other hand, any attempt to prepare PLA dispersions in the presence of higher MFC content was not successful because the mixtures could not be homogenized under ultrasound treatment due to the extremely high viscosity of the water phase.

Similar viscosity constraints were also observed with XG. Indeed, tests with XG as stabilizer provided stable dispersions only at XG concentration in water of 0.009 wt.% (entry 5, Table 2). The high viscosity of more concentrated XG solution, even 0.1 wt.%, prevented the effective emulsification of oil-and-water phases. Dispersion with 0.009 wt.% concentration of XG has nanoparticles with 192 nm diameter and dry residual of 14%. Both values are comparable to the ones afforded with starch as stabilizer, thus suggesting comparable mechanisms of stabilization for the two polysaccharides. Unluckily, when the preparation was performed at a larger scale (3 times higher) with XG, coagulation occurred (entry 6, Table 2). Overall, a PLA dispersion in water is formed in all the conditions listed in Table 2. However, only with starch as stabilizer is the formulation stable over time and it maintains its stability even when it is prepared at a higher scale, thus highlighting the key role of this polysaccharide in efficiently stabilizing the preparations, without preventing the effective homogenization of phases during the emulsion preparation.

Furthermore, the good stability of dispersions with starch, in spite of the lower viscosity of starch solutions with respect to XG, suggests that the viscosity is not the parameter that controls the dispersion stability.

3.4. Screening of Different Emulsifiers for the Formation and Stability of the Dispersion

In order to study the role of the emulsion stabilizer used in the formation and stability of the preparations, several surfactants in different combinations and concentrations were tested (Table 3).

At first, SDS in combination with Span 80, SYN (a PEO-PPO block copolymer), alone and in combination with Span 80 and Tween 80 were tested in the absence of starch (from entry 2 to 9, Table 3). However, only SDS in combination with Span 80 (entry 2, Table 3) and SYN at 2.4% concentration (entry 4, Table 3) provided stable emulsions. In the case of SYN, an almost satisfactory emulsion was already obtained at 0.8% (entry 6, Table 3). However, PLA coagulated on the stirrer forming a clot during the subsequent solvent evaporation, when the concentration was other than 2.4% (entries 3, 5, 6 and 7, Table 3). The dispersion obtained with SYN possesses a particle size between 400 and 450 nm. The value is higher than the one obtained with SDS. On the contrary, the particle size observed with SDS and Span 80 was around 150 nm, which is comparable to the one with only SDS.

Met. 388, SA, NaCMC, SSL and Met. 368 (entries 12, 15, 16, 17 and 18, Table 3) tested in the presence of starch, all led to the formation of inverted emulsions (water in oil instead of oil in water). On the other hand, with the use of Tween 80 and Tween 20 (entries 9, 13 and 14, Table 3) all the preparations were unstable due to coagulation.

In order to rationalize the obtained results, they are visualized in Figure 3: the formation and stability of the PLA preparations are plotted as a function of the surfactants' hydrophilic–lipophilic balance (HLB) values and their concentration in the water phase. The data highlight that for relatively low values of HLB (between 5 and 11) and low concentration in water (from 0.5 to 0.1 wt.%), an inverted emulsion (water-in-oil) is formed (red symbols, Figure 3). When the hydrophilic–lipophilic balance (HLB) value is in the 15–20 range, without exceeding concentration values of 0.1 wt.% in water, the emulsion sometimes is still inverted and sometimes partially forms (respectively, red and orange symbols, Figure 3). However, in this HLB range it is sufficient to increase the concentration of surfactant from 0.8 to 1.3 wt.% to obtain a direct emulsion (oil-in-water) (light green symbols, Figure 4). However, emulsions are not stable over time. Indeed, for hydrophilic–lipophilic balance (HLB) values higher than 19 and a concentration of surfactant higher than 1 wt.%, a direct emulsion was obtained with all tested surfactants, even if it is stable only in a few cases. In summary, two stability zones can be highlighted from the chart (dark green symbols, Figure 3): (a) for a high HLB value (40), a relatively low concentrations of surfactant (0.8 wt.%) and starch as stabilizer; and (b) with modest HLB (29 and 34) and fairly high concentrations of emulsifier (2 and 3 wt.%). In the second case, the use of starch as stabilizer does not affect the stability of the preparations, suggesting that, in a certain range of HLB values and concentrations of emulsifier, no stabilizing agents are needed. As a drawback, the use of a higher amount of surfactant significantly affects the composition of the final coating, therefore most likely also the final properties of the layer.

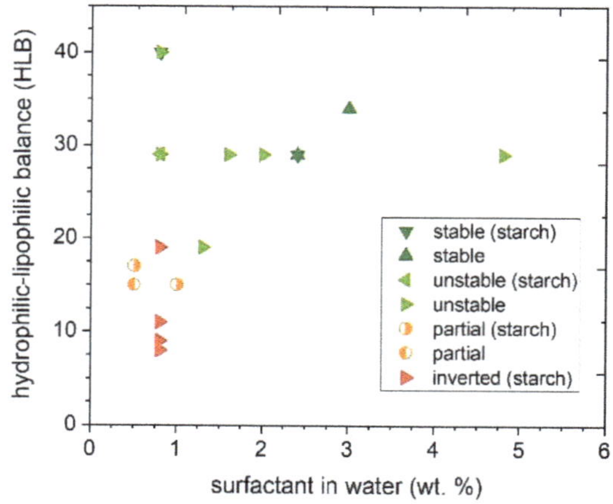

Figure 3. Influence of hydrophilic–lipophilic balance (HLB) and concentration of surfactant in water on the formation and stability of the formulations.

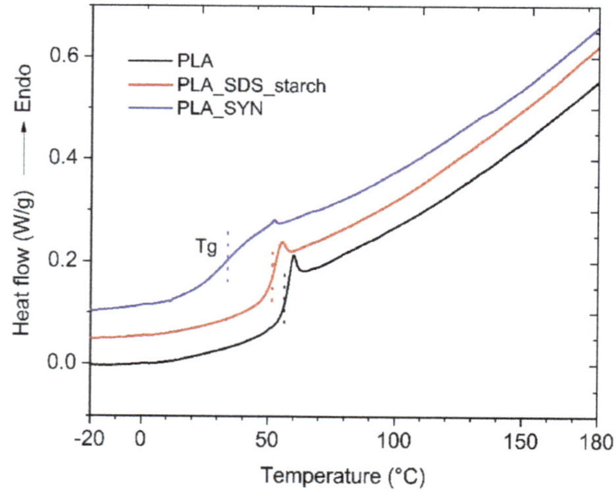

Figure 4. DSC thermograms at 10 °C/min of neat PLA, dried PLA dispersion with SDS and starch as surfactant and stabilizer, respectively, (entry 3, Table 1) and dried PLA dispersion with SYN (entry 23, Table 3). Second heating step. Thermograms were arbitrarily vertically shifted for clarity.

3.5. Characterization of the Dispersions

The dried dispersions were characterized through gel permeation chromatography (GPC) analysis in order to verify if the PLA was subjected to degradation during the preparation of the formulation and the subsequent storage. The comparison between the molecular weight of pristine PLA and PLA of dispersions showed minor differences (3–5%) in the range of the experimental error (Table 4). Furthermore, no significant change in polymer dispersity was detected, thus indicating the absence of significant degradation of the polyester during the processing in water for the dispersion preparation under ultrasound treatment and moderate heating and the subsequent storage for a few months.

DSC analysis of neat PLA 4060D showed glass transition temperature at 54.1 °C and no melting or crystallization peaks (Figure 4), in agreement with the amorphous nature of the resin declared by the producer (NatureWork processing guide NWPG002_020111; https://doi.org/10.1016/j.reactfunctpolym.2017.06.013, accessed on 1 September 2020).

The analysis of dried dispersion with SDS and starch (entry 3, Table 1) showed only a modest shift of the Tg value towards a lower value (Tg = 51 °C) with respect to the pristine polyester, thus indicating a modest effect of the emulsion process even on the thermal properties of the resin. On the contrary, dispersion with SYN showed a significant decrease of ~20 °C in the glass transition temperature (Tg = 34 °C), indicating a plasticization of PLA by the PEO block of SYN [78]. This latter sample, in addition to the shift in the Tg's mean value, showed a broadening of the transition, thus reflecting the variety of thermal motions available [79]. This last effect can be due to a non-homogeneous distribution of SYN in the PLA particles or even to a non-homogeneous distribution of SYN among particles, resulting in particles richer in the block copolymer and others with comparable depletion.

Table 4. Molecular weight by GPC analysis of pristine PLA and PLA dispersions dried at different temperatures.

Sample	Drying Temperature (°C)	$\overline{M_n}$ (kDa)	$\overline{M_w}$ (kDa)	Đ
PLA	-	113.8	166.1	1.46
PLA_SDS_starch	rt	109.6	157.8	1.44
PLA_SYN	rt	119.5	164.9	1.38
PLA_SDS_starch	60	96.8	141.3	1.46
PLA_SYN	60	102.2	134.9	1.32
PLA_SDS_starch	80	92.3	133.8	1.45
PLA_SYN	80	85.6	123.3	1.44
PLA_SDS_starch	100	99.6	143.4	1.44
PLA_SYN	100	88.9	131.5	1.48

3.6. Preparation of Films from PLA Dispersions

PLA dispersions in water prepared with either SDS_starch (entry 3, Table 1) or SYN (entry 29, Table 3) as surfactant and stabilizer were cast on PTFE capsules and left to dry. When drying was performed at room temperature, a layer of powder was obtained. SEM analysis revealed the presence of submicrometric particles (Figure 5a,b), whose dimension was comparable to the values by DLS data of the corresponding water dispersion (Section 3.2). This result indicates that particles did not merge, most likely because the processing temperature was lower than the glass transition value of both dispersions (Section 3.5). When drying was performed at 40 °C, particles started to merge, forming white continuous layers. However, these were brittle and fractured (Figure 6a,b). A further increase of the temperature to 50 °C had a partial positive effect only in the case of the SYN dispersion, which has a glass transition temperature of 34 °C. However, the formed film, even if partially transparent, was brittle and cracked. On the contrary, when films are dried at 60 °C, continuous free-standing films with thickness between 230 and 550 μm are afforded (Figure 6c,d) with both formulations.

The formulation with SYN gave even, almost transparent films, while in the presence of SDS and starch, opaque films are afforded. In both cases, individual particles cannot be distinguished by SEM observation of the film's surfaces (Figure 5c,e,g,h) and section (Figure 5d,f not shown), thus indicating effective film formation. Notice that 60 °C is higher than the glass transition temperature of both dispersions and of the pristine PLA. Processing at higher temperatures, up to 110 °C, provided films that look like the one obtained at 60 °C in the case of the SYN formulation. However, with the dispersions with SDS and starch, yellowing occurred at drying temperatures above 90 °C. Indeed, the SEM analysis of PLA dispersion in water prepared with SDS_starch at 80 °C (Figure 5g) shows the presence of granules that are probably due to the degradation of the surfactant/stabilizer.

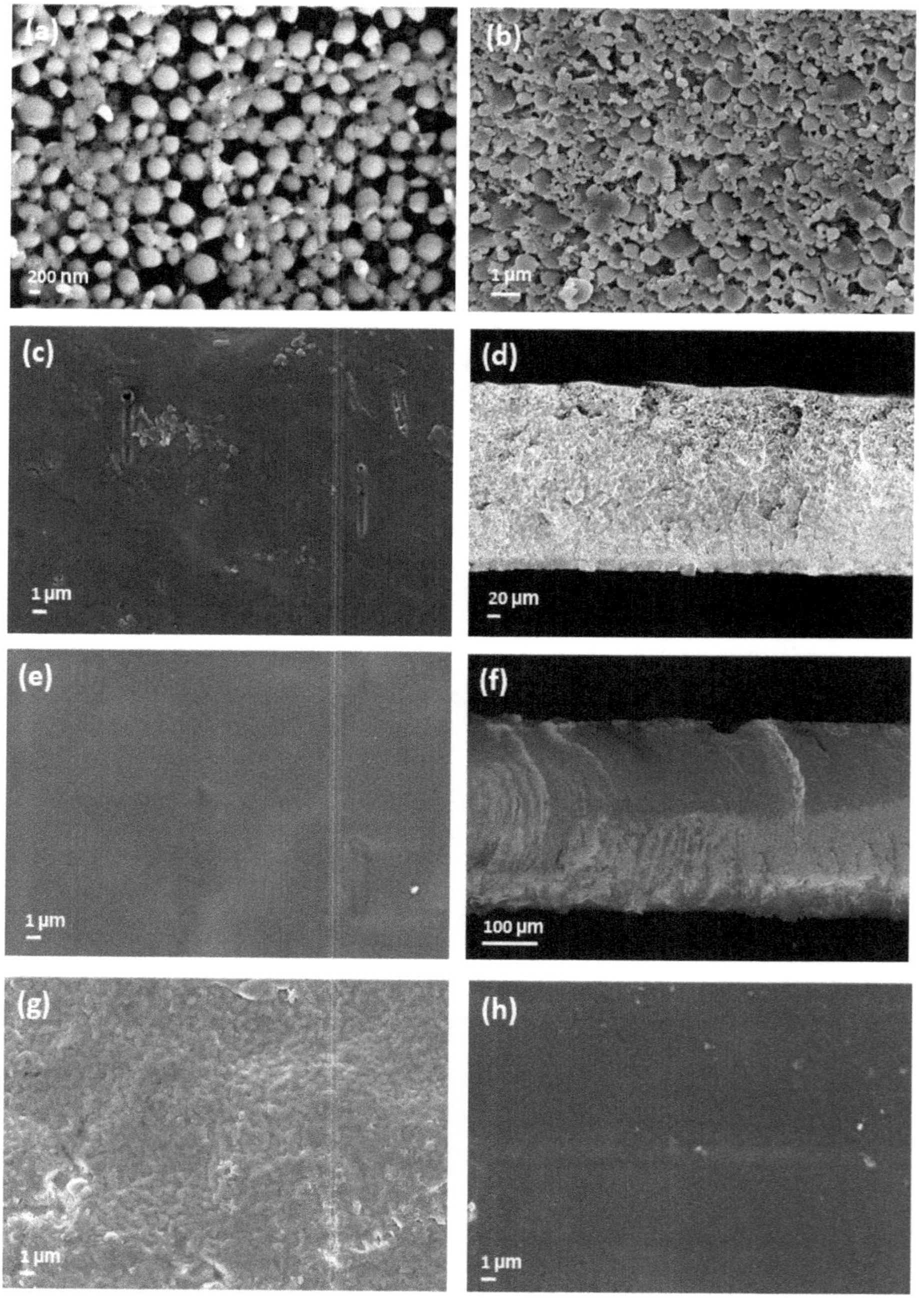

Figure 5. SEM pictures of the filming tests performed with water dispersions of PLA at room temperature (**a**,**b**), at 60 °C (**c**–**f**) and at 80 °C (**g**,**h**). Picture (**a**,**c**,**d**,**g**) refer to tests with the SDS_starch formulation; (**b**,**e**,**f**,**h**), correspond to tests with the SYN formulation.

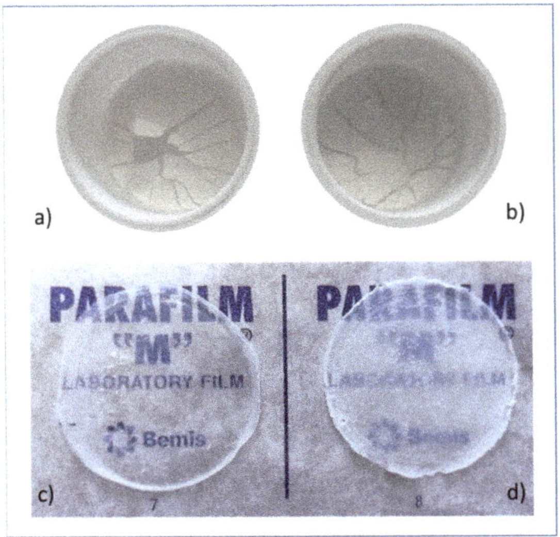

Figure 6. Pictures of PLA films obtained by solution casting at 40 °C (**a**,**b**) and 60 °C (**c**,**d**) of water dispersions with SDS and starch (**a**,**d**) or SYN (**b**,**c**).

Transmission measurements showed that PLA films with SDS and starch (thickness of 320 µm) possess a transmittance of 2.6% at 600 nm. The low transmittance is due to the whitish color of the film, probably given by the presence of starch. On the other hand, PLA film with SYN (thickness of 250 µm) showed a higher transparency (transmittance of 40% at 600 nm). However, the latter possesses a modest transparency considering that a pure PLA film prepared from EtOAc (thickness of 130 µm) has a transmittance of 72% at 600 nm.

ATR analysis of the upper and lower surfaces of the films showed comparable band intensity, hence indicating a similar composition. As a consequence, significant surfactant segregation or blooming effects can be excluded (Figure 7). However, films obtained at high drying temperatures (90 and 110 °C) showed a different composition between the upper and lower surface by ATR analysis (Figure 7). In particular, an enrichment in stabilizer/surfactant is observed on the lower surface of the films. Indeed, the IR spectra of the bottom surface of PLA_SDS_starch film prepared at 90 °C (Figure 7a) shows a broad band between 3000 and 3500 cm^{-1}, which is typical of the starch's OH stretching. The enrichment of starch is further confirmed by the typical stretching band of sugar rings at around 1000 cm^{-1}. Similarly, the film from the PLA_SYN formulation obtained at 110 °C, in the bottom surface showed more intense signal than the top surface. These are signals at nearly 2800 cm^{-1} relative to CH_2 stretching and the one at nearly 955 cm^{-1}, which are both present in the ATR spectra of Synperonic.

The possible degradation of PLA during film formation was checked by GPC analysis (Table 4). A modest variation, even if detectable (10–15% decrement), was observed after drying at 60 °C. Higher processing temperature resulted in a more extended molecular weight decrement. No variation of the dispersity was detected as a result of a comparable decrement in both the $\overline{M_n}$ and $\overline{M_w}$ value, thus suggesting no preferential degradation of the high or low molecular weight fractions. In any case, no optimization of the drying time has been performed and it can be envisaged that better control over the molecular weight can be achieved by proper control of the heating time.

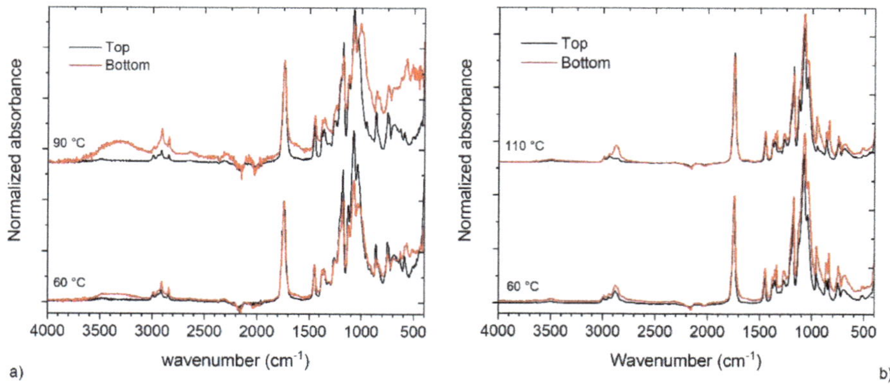

Figure 7. (**a**) FT-IR spectra of the top and bottom surfaces of PLA_SDS_starch films prepared at 60 and 90 °C; (**b**) FT-IR spectra of the top and bottom surfaces of PLA_SYN film prepared at 60 and 110 °C.

4. Conclusions

In the present work, a procedure based on emulsification-evaporation method, which gives PLA dispersions in water with dry matter content up to 20 wt.%. has been developed. The procedure is based on the combination of an amorphous commercial PLA grade, ethyl acetate as solvent and surfactants all approved for food contact applications, so that the new preparation is also suitable for this application's purpose.

Surfactants with hydrophilic–lipophilic balance (HLB) in the 29–34 range at 2 and 3 wt.% concentration are needed to stabilize the formulations during the preparation stages. Surfactants with higher HLB (40) are also effective if starch is included as stabilizer. Emulsification needs high energy input by ultrasound sonication to achieve effective dispersion. Furthermore, the total treated volumes, the organic/water volume ratio and the viscosity of phases must be adjusted to the sonication capability to achieve the stability of the final dispersion. In addition, for long-term stability, the final dispersion formulation must contain negligible ethyl acetate residua (<40 ppm). Under this circumstance, dispersions do not sedimentate or cream and particle size does not appreciably change over several months at room temperature or in the fridge at 4 °C.

The obtained dispersions, consisting of submicrometric PLA particles, were successfully cast into films, with minimum formation temperature of 60 °C, which is a little above the glass transition temperature of the pristine PLA.

5. Patents

The method of preparing halogen-free aqueous dispersion of biodegradable polymers was submitted to the Italian Patent Office (Bertoldo, M.; Ricci, L.; Messina, T.; Belletti, G.; Gallur Blanca, M.; Guillem Ortiz, A. and Aragón Gutierrez, A. Italian Patent issue no. 102020000028640, 26 November 2020).

Author Contributions: Conceptualization, M.B.; methodology, M.B., G.B. and L.R.; validation, G.B. and S.B.; formal analysis, G.B. and S.B.; investigation, G.B. and S.B.; resources, M.B. and O.B.; data curation, M.B. and G.B.; writing—original draft preparation, M.B. and G.B.; writing—review and editing, M.B., G.B., A.G.-O. and A.A.-G.; visualization, M.B., G.B., A.G.-O. and A.A.-G.; supervision, M.B. and O.B.; project administration, M.B.; funding acquisition, M.B. All authors have read and agreed to the published version of the manuscript.

Funding: This research was funded by the Bio Based Industries Joint Undertaking under the European Union's Horizon 2020 research and innovation program under grant agreement No 745718 (Sherpack).

Institutional Review Board Statement: Not applicable.

Informed Consent Statement: Not applicable.

Data Availability Statement: The data presented in this study are available on request from the corresponding author.

Acknowledgments: Professor Valter Castelvetro of the University of Pisa is kindly acknowledged for the GPC analysis. Franco Corticelli of Institute for Microelectronics and Microsystem (National Research Council) is kindly acknowledged for the SEM analysis. Cargill Spa is acknowledged for providing Starch C*IcoatTM 07525 and xanthan gum, Borregaard for MFC Exilva F 01-L and Croda Chemicals for Synperonic PE/F68.

Conflicts of Interest: The authors declare no conflict of interest.

References

1. Rhim, J.-W.; Lee, J.-H.; Hong, S.-I. Increase in water resistance of paperboard by coating with poly(lactide). *Packag. Technol. Sci.* **2007**, *20*, 393–402. [CrossRef]
2. Farah, S.; Anderson, D.G.; Langer, R. Physical and mechanical properties of PLA, and their functions in widespread applications—A comprehensive review. *Adv. Drug Deliv. Rev.* **2016**, *107*, 367–392. [CrossRef]
3. Habel, C.; Schöttle, M.; Daab, M.; Eichstaedt, N.J.; Wagner, D.; Bakhshi, H.; Agarwal, S.; Horn, M.A.; Breu, J. High-Barrier, Biodegradable Food Packaging. *Macromol. Mater. Eng.* **2018**, *303*, 1800333. [CrossRef]
4. Auras, R.; Harte, B.; Selke, S. An Overview of Polylactides as Packaging Materials. *Macromol. Biosci.* **2004**, *4*, 835–864. [CrossRef] [PubMed]
5. Tawakkal, I.S.M.A.; Cran, M.J.; Miltz, J.; Bigger, S. A Review of Poly(Lactic Acid)-Based Materials for Antimicrobial Packaging. *J. Food Sci.* **2014**, *79*, R1477–R1490. [CrossRef] [PubMed]
6. Ahmed, J.; Varshney, S.K. Polylactides—Chemistry, Properties and Green Packaging Technology: A Review. *Int. J. Food Prop.* **2011**, *14*, 37–58. [CrossRef]
7. Jabeen, N.; Majid, I.; Nayik, G.A. Bioplastics and food packaging: A review. *Cogent Food Agric.* **2015**, *1*, 1117749. [CrossRef]
8. Lim, L.-T.; Auras, R.; Rubino, M. Processing technologies for poly(lactic acid). *Prog. Polym. Sci.* **2008**, *33*, 820–852. [CrossRef]
9. Castro-Aguirre, E.; Iñiguez-Franco, F.; Samsudin, H.; Fang, X.; Auras, R. Poly(lactic acid)—Mass production, processing, industrial applications, and end of life. *Adv. Drug Deliv. Rev.* **2016**, *107*, 333–366. [CrossRef]
10. Rhim, J.-W.; Mohanty, A.K.; Singh, S.P.; Ng, P.K.W. Effect of the processing methods on the performance of polylactide films: Thermocompression versus solvent casting. *J. Appl. Polym. Sci.* **2006**, *101*, 3736–3742. [CrossRef]
11. Wang, L.-F.; Rhim, J.-W.; Hong, S.-I. Preparation of poly(lactide)/poly(butylene adipate-co-terephthalate) blend films using a solvent casting method and their food packaging application. *LWT* **2016**, *68*, 454–461. [CrossRef]
12. Duan, H.; Shao, Z.; Zhao, M.; Zhou, Z. Preparation and properties of environmental-friendly coatings based on carboxymethyl cellulose nitrate ester & modified alkyd. *Carbohydr. Polym.* **2016**, *137*, 92–99. [CrossRef]
13. Athawale, V.D.; Nimbalkar, R.V. Waterborne Coatings Based on Renewable Oil Resources: An Overview. *J. Am. Oil Chem. Soc.* **2010**, *88*, 159–185. [CrossRef]
14. Horváth, I.T. Solvents from nature. *Green Chem.* **2008**, *10*, 1024–1028. [CrossRef]
15. Mehravar, S.; Ballard, N.; Tomovska, R.; Asua, J.M. Polyurethane/Acrylic Hybrid Waterborne Dispersions: Synthesis, Properties and Applications. *Ind. Eng. Chem. Res.* **2019**, *58*, 20902–20922. [CrossRef]
16. Lossada, F.; Jiao, D.; Yao, X.; Walther, A. Waterborne Methacrylate-Based Vitrimers. *ACS Macro Lett.* **2019**, *9*, 70–76. [CrossRef]
17. Asif, A.; Hu, L.; Shi, W. Synthesis, rheological, and thermal properties of waterborne hyperbranched polyurethane acrylate dispersions for UV curable coatings. *Colloid Polym. Sci.* **2009**, *287*, 1041–1049. [CrossRef]
18. Santamaria-Echart, A.; Fernandes, I.; Barreiro, F.; Corcuera, M.; Eceiza, A. Advances in Waterborne Polyurethane and Polyurethane-Urea Dispersions and Their Eco-friendly Derivatives: A Review. *Polymers* **2021**, *13*, 409. [CrossRef]
19. Alvarez, G.A.; Fuensanta, M.; Orozco, V.; Giraldo, L.F.; Martín-Martínez, J.M. Hybrid waterborne polyurethane/acrylate dispersion synthesized with bisphenol A-glicidylmethacrylate (Bis-GMA) grafting agent. *Prog. Org. Coat.* **2018**, *118*, 30–39. [CrossRef]
20. Zhang, T.; Wu, W.; Wang, X.; Mu, Y. Effect of average functionality on properties of UV-curable waterborne polyurethane-acrylate. *Prog. Org. Coat.* **2010**, *68*, 201–207. [CrossRef]
21. Diaconu, G.; Paulis, M.; Leiza, J.R. Towards the synthesis of high solids content waterborne poly(methyl methacrylate-co-butyl acrylate)/montmorillonite nanocomposites. *Polymers* **2008**, *49*, 2444–2454. [CrossRef]
22. Guo, X.; Ge, S.; Wang, J.; Zhang, X.; Zhang, T.; Lin, J.; Zhao, C.X.; Wang, B.; Zhu, G.; Guo, Z. Waterborne acrylic resin modified with glycidyl methacrylate (GMA): Formula optimization and property analysis. *Polymers* **2018**, *143*, 155–163. [CrossRef]
23. Di Crescenzo, M.M.; Zendri, E.; Sánchez-Pons, M.; Fuster-López, L.; Yusá-Marco, D.J. The use of waterborne paints in contemporary murals: Comparing the stability of vinyl, acrylic and styrene-acrylic formulations to outdoor weathering conditions. *Polym. Degrad. Stab.* **2014**, *107*, 285–293. [CrossRef]
24. Grøstad, K.; Pedersen, A. Emulsion Polymer Isocyanates as Wood Adhesive: A Review. *J. Adhes. Sci. Technol.* **2010**, *24*, 1357–1381. [CrossRef]
25. Zhou, X.; Li, Y.; Fang, C.; Li, S.; Cheng, Y.; Lei, W.; Meng, X. Recent Advances in Synthesis of Waterborne Polyurethane and Their Application in Water-based Ink: A Review. *J. Mater. Sci. Technol.* **2015**, *31*, 708–722. [CrossRef]

26. Lei, L.; Zhong, L.; Lin, X.; Li, Y.; Xia, Z. Synthesis and characterization of waterborne polyurethane dispersions with different chain extenders for potential application in waterborne ink. *Chem. Eng. J.* **2014**, *253*, 518–525. [CrossRef]
27. Bandera, D.; Meyer, V.R.; Prevost, D.; Zimmermann, T.; Boesel, L.F. Polylactide/Montmorillonite Hybrid Latex as a Barrier Coating for Paper Applications. *Polymers* **2016**, *8*, 75. [CrossRef] [PubMed]
28. Nestorson, A.; Neoh, K.G.; Kang, E.-T.; Järnström, L.; Leufvén, A. Enzyme immobilization in latex dispersion coatings for active food packaging. *Packag. Technol. Sci.* **2008**, *21*, 193–205. [CrossRef]
29. Imoto, T. Film formation mechanism of latices. *J. Jpn. Soc. Colour Mater.* **1966**, *39*, 34–41. [CrossRef]
30. Felton, L.A. Mechanisms of polymeric film formation. *Int. J. Pharm.* **2013**, *457*, 423–427. [CrossRef]
31. Keddie, J.; Meredith, P.; Jones, R.; Donald, A.M. Kinetics of Film Formation in Acrylic Latices Studied with Multiple-Angle-of-Incidence Ellipsometry and Environmental SEM. *Macromol.* **1995**, *28*, 2673–2682. [CrossRef]
32. Cummings, S.; Zhang, Y.; Smeets, N.; Cunningham, M.; Dubé, M.A. On the Use of Starch in Emulsion Polymerizations. *Process* **2019**, *7*, 140. [CrossRef]
33. Keddie, J.L. Reports: A Review Journal Film formation of latex. *Mater. Sci. Eng.* **1997**, *21*, 101–170. [CrossRef]
34. Brodnyan, J.G.; Cala, J.A.; Konen, T.; Kelley, E. The mechanism of emulsion polymerization. I. Studies of the polymerization of methyl methacrylate and n-butyl methacrylate. *J. Colloid Sci.* **1963**, *18*, 73–90. [CrossRef]
35. Bao, J.; Zhang, A. Poly(methyl methacrylate) nanoparticles prepared through microwave emulsion polymerization. *J. Appl. Polym. Sci.* **2004**, *93*, 2815–2820. [CrossRef]
36. Meneghetti, P.; Qutubuddin, S. Synthesis of Poly(methyl methacrylate) Nanocomposites via Emulsion Polymerization Using a Zwitterionic Surfactant. *Langmuir* **2004**, *20*, 3424–3430. [CrossRef]
37. Qiu, J.; Gaynor, S.G.; Matyjaszewski, K. Emulsion Polymerization of n-Butyl Methacrylate by Reverse Atom Transfer Radical Polymerization. *Macromolecules* **1999**, *32*, 2872–2875. [CrossRef]
38. Xaviera, J.L.L.; Guyota, A.; Bourgeat-Lami, E. Synthesis and Characterization of Silica/Poly (Methyl Methacrylate) Nanocomposite Latex Particles through Emulsion Polymerization Using a Cationic Azo Initiator. *J. Colloid Interface Sci.* **2002**, *250*, 82–92. [CrossRef]
39. Noreen, A.; Zia, K.M.; Zuber, M.; Tabasum, S.; Saif, M.J. Recent trends in environmentally friendly water-borne polyurethane coatings: A review. *Korean J. Chem. Eng.* **2015**, *33*, 388–400. [CrossRef]
40. Agnol, L.D.; Dias, F.T.G.; Ornaghi, H.L.; Sangermano, M.; Bianchi, O. UV-curable waterborne polyurethane coatings: A state-of-the-art and recent advances review. *Prog. Org. Coat.* **2021**, *154*, 106156. [CrossRef]
41. Allémann, E.; Gurny, R.; Doelker, E. Preparation of aqueous polymeric nanodispersions by a reversible salting-out process: Influence of process parameters on particle size. *Int. J. Pharm.* **1992**, *87*, 247–253. [CrossRef]
42. Grunlan, M.A.; Xing, L.-L.; Glass, J.E. Waterborne Coatings with an Emphasis on Synthetic Aspects: An Overview. *ACS Symp. Ser.* **1997**, 1–26. [CrossRef]
43. Noble, K.-L. Waterborne polyurethanes. *Prog. Org. Coat.* **1997**, *32*, 131–136. [CrossRef]
44. Lassalle, V.; Ferreira, M.L. PLA Nano- and Microparticles for Drug Delivery: An Overview of the Methods of Preparation. *Macromol. Biosci.* **2007**, *7*, 767–783. [CrossRef] [PubMed]
45. Lee, B.K.; Yun, Y.; Park, K. PLA micro- and nano-particles. *Adv. Drug Deliv. Rev.* **2016**, *107*, 176–191. [CrossRef]
46. Hong, J.S.; Srivastava, D.; Lee, I. Fabrication of poly(lactic acid) nano- and microparticles using a nanomixer via nanoprecipitation or emulsion diffusion. *J. Appl. Polym. Sci.* **2018**, *135*, 46199. [CrossRef]
47. Legrand, P.; Lesieur, S.; Bochot, A.; Gref, R.; Raatjes, W.; Barratt, G.; Vauthier, C. Influence of polymer behaviour in organic solution on the production of polylactide nanoparticles by nanoprecipitation. *Int. J. Pharm.* **2007**, *344*, 33–43. [CrossRef] [PubMed]
48. Pandey, S.K.; Patel, D.; Thakur, R.; Mishra, D.P.; Maiti, P.; Haldar, C. Anti-cancer evaluation of quercetin embedded PLA nanoparticles synthesized by emulsified nanoprecipitation. *Int. J. Biol. Macromol.* **2015**, *75*, 521–529. [CrossRef]
49. Gavory, C.; Durand, A.; Six, J.-L.; Nouvel, C.; Marie, E.; Leonard, M. Polysaccharide-covered nanoparticles prepared by nanoprecipitation. *Carbohydr. Polym.* **2011**, *84*, 133–140. [CrossRef]
50. Allémann, E.; Leroux, J.; Gurny, R.; Doelker, E. In Vitro Extended-Release Properties of Drug-Loaded Poly(DL-Lactic Acid) Nanoparticles Produced by a Salting-Out Procedure. *Pharm. Res.* **1993**, *10*, 1732–1737. [CrossRef]
51. Mendoza-Muñoz, N.; Quintanar-Guerrero, D.; Allémann, E. The impact of the salting-out technique on the preparation of colloidal particulate systems for pharmaceutical applications. *Recent Patents Drug Deliv. Formul.* **2012**, *6*, 236–249. [CrossRef] [PubMed]
52. Balakrishanan, M.H.; Rajan, M. Size-controlled synthesis of biodegradable nanocarriers for targeted and controlled cancer drug delivery using salting out cation. *Bull. Mater. Sci.* **2016**, *39*, 69–77. [CrossRef]
53. Trimaille, T.; Pichot, C.; Fessi, H.; Delair, T. Poly(d,l-lactic acid) nanoparticle preparation and colloidal characterization. *Colloid Polym. Sci.* **2003**, *281*, 1184–1190. [CrossRef]
54. Quintanar-Guerrero, D.; Fessi, H.; Allémann, E.; Doelker, E. Influence of stabilizing agents and preparative variables on the formation of poly(d,l-lactic acid) nanoparticles by an emulsification-diffusion technique. *Int. J. Pharm.* **1996**, *143*, 133–141. [CrossRef]
55. Coffin, M.D.; McGinity, J.W. Biodegradable Pseudolatexes: The Chemical Stability of Poly(D,L-Lactide) and Poly (ε-Caprolactone) Nanoparticles in Aqueous Media. *Pharm. Res. Off. J. Am. Assoc. Pharm. Sci.* **1992**, *9*, 200–205.

56. Zhang, H.; Xiang, S.; Luan, Q.; Bao, Y.; Deng, Q.; Zheng, M.; Liu, S.; Song, J.; Tang, H.; Huang, F. Development of poly (lactic acid) microspheres and their potential application in Pickering emulsions stabilization. *Int. J. Biol. Macromol.* **2018**, *108*, 105–111. [CrossRef]
57. Xu, W.; Shen, R.; Yan, Y.; Gao, J. Preparation and characterization of electrospun alginate/PLA nanofibers as tissue engineering material by emulsion eletrospinning. *J. Mech. Behav. Biomed. Mater.* **2017**, *65*, 428–438. [CrossRef]
58. Sébastien, F.; Stéphane, G.; Copinet, A.; Coma, V. Novel biodegradable films made from chitosan and poly(lactic acid) with antifungal properties against mycotoxinogen strains. *Carbohydr. Polym.* **2006**, *65*, 185–193. [CrossRef]
59. Grande, R.; Carvalho, A. Compatible Ternary Blends of Chitosan/poly(vinyl alcohol)/poly(lactic acid) Produced by Oil-in-Water Emulsion Processing. *Biomacromolecules* **2011**, *12*, 907–914. [CrossRef] [PubMed]
60. Li, X.; Hegyesi, N.; Zhang, Y.; Mao, Z.; Feng, X.; Wang, B.; Pukánszky, B.; Sui, X. Poly(lactic acid)/lignin blends prepared with the Pickering emulsion template method. *Eur. Polym. J.* **2018**, *110*, 378–384. [CrossRef]
61. Zhang, Y.; Wu, J.; Wang, B.; Sui, X.; Zhong, Y.; Zhang, L.; Mao, Z.; Xu, H. Cellulose nanofibril-reinforced biodegradable polymer composites obtained via a Pickering emulsion approach. *Cellulose* **2017**, *24*, 3313–3322. [CrossRef]
62. Zhang, Y.; Jiang, Y.; Han, L.; Wang, B.; Xu, H.; Zhong, Y.; Zhang, L.; Mao, Z.; Sui, X. Biodegradable regenerated cellulose-dispersed composites with improved properties via a pickering emulsion process. *Carbohydr. Polym.* **2018**, *179*, 86–92. [CrossRef] [PubMed]
63. Horianopoulos, D.; Horianopoulos, S. Traffic-Actuated Electrical Generator Apparatus. U.S. Patent 20070085342A1, 19 April 2007.
64. Kratz, F. Bisphosphonate-Prodrugs. Patent No. WO/2011/023367, 3 March 2011.
65. Zembyla, M.; Murray, B.S.; Sarkar, A. Water-in-oil emulsions stabilized by surfactants, biopolymers and/or particles: A review. *Trends Food Sci. Technol.* **2020**, *104*, 49–59. [CrossRef]
66. Kume, G.; Gallotti, M.; Nunes, G. Review on Anionic/Cationic Surfactant Mixtures. *J. Surfactants Deterg.* **2007**, *11*, 1–11. [CrossRef]
67. Foley, P.; Pour, A.K.; Beach, E.; Zimmerman, J. Derivation and synthesis of renewable surfactants. *Chem. Soc. Rev.* **2011**, *41*, 1499–1518. [CrossRef] [PubMed]
68. Dickinson, E. Mixed biopolymers at interfaces: Competitive adsorption and multilayer structures. *Food Hydrocoll.* **2011**, *25*, 1966–1983. [CrossRef]
69. Li, C.; Fu, X.; Luo, F.; Huang, Q. Effects of maltose on stability and rheological properties of orange oil-in-water emulsion formed by OSA modified starch. *Food Hydrocoll.* **2012**, *32*, 79–86. [CrossRef]
70. Pasquali, R.C.; Taurozzi, M.P.; Bregni, C. Some considerations about the hydrophilic–lipophilic balance system. *Int. J. Pharm.* **2008**, *356*, 44–51. [CrossRef] [PubMed]
71. Habibi, H.; Khosravi-Darani, K. Effective variables on production and structure of xanthan gum and its food applications: A review. *Biocatal. Agric. Biotechnol.* **2017**, *10*, 130–140. [CrossRef]
72. Yazid, N.S.M.; Abdullah, N.; Muhammad, N.; Matias-Peralta, H.M. Application of Starch and Starch-Based Products in Food Industry. *J. Sci. Technol.* **2018**, *10*. [CrossRef]
73. Winuprasith, T.; Suphantharika, M. Microfibrillated cellulose from mangosteen (Garcinia mangostana L.) rind: Preparation, characterization, and evaluation as an emulsion stabilizer. *Food Hydrocoll.* **2013**, *32*, 383–394. [CrossRef]
74. Raso, J.; Mañas, P.; Pagán, R.; Sala, F.J. Influence of different factors on the output power transferred into medium by ultrasound. *Ultrason. Sonochem.* **1999**, *5*, 157–162. [CrossRef]
75. Agama-Acevedo, E.; Bello-Perez, L. Starch as an emulsions stability: The case of octenyl succinic anhydride (OSA) starch. *Curr. Opin. Food Sci.* **2017**, *13*, 78–83. [CrossRef]
76. Winuprasith, T.; Suphantharika, M. Properties and stability of oil-in-water emulsions stabilized by microfibrillated cellulose from mangosteen rind. *Food Hydrocoll.* **2015**, *43*, 690–699. [CrossRef]
77. Hemar, Y.; Tamehana, M.; Munro, P.; Singh, H. Influence of xanthan gum on the formation and stability of sodium caseinate oil-in-water emulsions. *Food Hydrocoll.* **2001**, *15*, 513–519. [CrossRef]
78. Eom, Y.; Choi, B.; Park, S.-I. A Study on Mechanical and Thermal Properties of PLA/PEO Blends. *J. Polym. Environ.* **2018**, *27*, 256–262. [CrossRef]
79. Erukhimovich, I.; de la Cruz, M.O. Phase equilibria and charge fractionation in polydisperse polyelectrolyte solutions. *J. Polym. Sci. Part B: Polym. Phys.* **2007**, *45*, 3003–3009. [CrossRef]

MDPI
St. Alban-Anlage 66
4052 Basel
Switzerland
Tel. +41 61 683 77 34
Fax +41 61 302 89 18
www.mdpi.com

Polymers Editorial Office
E-mail: polymers@mdpi.com
www.mdpi.com/journal/polymers

www.ingramcontent.com/pod-product-compliance
Lightning Source LLC
Chambersburg PA
CBHW062103220926
51850CB00010B/2868